SCIENCE IN EUROPE, 1500–1800
A PRIMARY SOURCES READER

Science in Europe, 1500–1800: A Primary Sources Reader

The companion volume in this series is:

Science in Europe, 1500–1800: A Secondary Sources Reader

Both volumes are part of an Open University course, *The Rise of Scientific Europe 1500–1800* (AS208), a 60 points second level undergraduate course.

Opinions expressed in this Reader are not necessarily those of the Course Team or of The Open University.

Details of this and other Open University courses can be obtained from the Call Centre, PO Box 724, The Open University, Milton Keynes, MK7 6ZS, United Kingdom: tel: +44 (0)1908 653231; e-mail: ces-gen@open.ac.uk.

Alternatively, you may wish to visit the Open University website at http://www.open.ac.uk where you can learn about the wide range of courses and packs offered at all levels by The Open University.

For information about the purchase of Open University course components, contact Open University Worldwide Ltd, The Berrill Building, Walton Hall, Milton Keynes, MK7 6AA, United Kingdom: tel: +44 (0)1908 858785; fax: +44 (0)1908 858787; e-mail: ouwenq@open.ac.uk; website: http://www.ouw.co.uk.

SCIENCE IN EUROPE, 1500–1800

A PRIMARY SOURCES READER

Edited by
Malcolm Oster

in association with

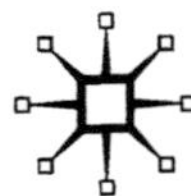

First published 2002 by
Palgrave in association with The Open University

PALGRAVE MACMILLAN
Houndmills, Basingstoke, Hampshire RG21 6XS and
175 Fifth Avenue, New York, N.Y. 10010
Companies and representatives throughout the world

PALGRAVE MACMILLAN is the new global academic imprint of St. Martin's Press LLC Scholarly and Reference Division and Palgrave Publishers Ltd (formerly Macmillan Press Ltd).

ISBN-13: 978–0–333–97002–7 paperback

This book is printed on paper suitable for recycling and made from fully managed and sustained forest sources. Logging, pulping and manufacturing processes are expected to conform to the environmental regulations of the country of origin

A catalogue record for this book is available
from the British Library.

Library of Congress Cataloging-in-Publication Data
Science in Europe, 1500–1800 : a primary sources reader / edited by Malcolm Oster.
p. cm.
Includes bibliographical references and index.

1. Science—Europe—History—Sources. I. Oster, Malcolm, 1952–

Q127.E8 S352 2001
509.4′09′03—dc21 2001036833

Transferred to Digital Printing 2011

Contents

Dramatis Personae

Chapter 1: Europe's Awakening

1.1 Plato (427–347 BC), Greek philosopher
1.2 Aristotle (384–322 BC), Greek philosopher
1.3 Lucretius (*c.*95–55 BC), Roman philosopher
1.4, 1.5 Galen (AD 129–*c.*200), Greek physician and anatomist
1.6 Johannes de Sacrobosco (*fl.* 1230), Irish mathematician

Chapter 2: Copernicus and his Revolution

2.1 (a) (d) (e) Nicholas Copernicus (1473–1543), Polish astronomer and mathematician
2.1 (b) Andreas Osiander (1498–1552), German astronomer and theologian
2.1 (c) Nicholas Schönberg
2.2 Georgius Rheticus (1514–1574), German mathematician
2.3 Andreas Vesalius (1514–1564), Flemish physician and anatomist

Chapter 3: The Spread of Copernicanism in Northern Europe

3.1 Robert Recorde (1510–1558), English mathematical practitioner and physician
3.2 Edward Wright (1558–1615), English mathematical practitioner
3.3 Tycho Brahe (1546–1601), Danish astronomer
3.4, 3.5 Johannes Kepler (1571–1630), German astronomer and mathematician
3.6 John Wilkins (1614–1672), English mathematician and cleric

Chapter 4: Crisis in Italy

4.1 (a), 4.2, 4.3, 4.4 Galileo Galilei (1564–1642), Italian mathematician and natural philosopher

4.1 (b) Roberto Bellarmine (1542–1621), Jesuit philosopher and theologian
4.5 Tommasso Campanella (1568–1639), Italian Dominican

Chapter 5: Iberian Science: Navigation, Empire and Counter-Reformation

5.1 Garcia d'Orta (*c.*1501–1568), Portuguese physician
5.2 Nicolás Monardes (*c.*1493–1588), Spanish physician
5.3, 5.4 Diego de Zúñiga (*c.*1536–*c.*1600), Spanish theologian and philosopher
5.5 Gaspar de Quiroga (1509–1594), Spanish cardinal

Chapter 6: Science from the Earth in Central Europe

6.1 Ulrich Rülein von Calw (*fl.* 1500), German physician and mineralogist
6.2 Cornelius Agrippa (1486–1535), German physician and occultist
6.3, 6.4, 6.5 Georgius Agricola (1494–1555), German physician and mineralogist
6.6, 6.7, 6.8 Paracelsus (*c.*1493–1541), Swiss physician, medical reformer and alchemist
6.9 Oswald Croll (*c.*1560–1609), German Paracelsian physician and alchemist

Chapter 7: French Science in the Seventeenth Century

7.1 René Descartes (1596–1650), French philosopher and mathematician
7.2 Blaise Pascal (1623–1662), French philosopher and mathematician

Chapter 8: Science in Seventeenth-Century England

8.1 William Gilbert (1544–1603), English physician and natural philosopher
8.2 William Harvey (1578–1657), English physician and anatomist
8.3 Francis Bacon (1561–1626), English statesman, lawyer, essayist and philosopher
8.4 Robert Hooke (1635–1702), English experimentalist
8.5, 8.6, 8.7 Robert Boyle (1627–1691), Anglo-Irish chemist, experimentalist and natural philosopher
8.8, 8.9, 8.10 Isaac Newton (1642–1727), English mathematician and natural philosopher

Chapter 9: Scientific Academies across Europe

9.1 Lorenzo Magalotti (1627–1712), Italian virtuoso
9.2 Thomas Sprat (1635–1713), English cleric and virtuoso
9.3 Henry Oldenburg (*c.*1618–1677), German expatriate philosophical correspondent

Chapter 10: The Reception of Newtonianism across Europe

10.1 Richard Bentley (1662–1742), English cleric and natural theologian
10.2 Isaac Newton (1642–1727), English mathematician and natural philosopher
10.3 John Ray (1627–1705), English naturalist
10.4 Roger Cotes (1682–1716), English mathematician and physicist
10.5 (a) (c) (d) G. W. Leibniz (1646–1716), German philosopher and mathematician
10.5 (b) Isaac Newton (1642–1727), English mathematician and natural philosopher
10.6 W. J. 'sGravesande (1688–1742), Dutch mathematician and physicist

Chapter 11: Science in the Scottish Enlightenment

11.1, 11.2 James Hutton (1726–1797), Scottish geologist and naturalist
11.3 Joseph Black (1728–1799), Scottish chemist and physicist
11.4 Adam Smith (1723–1790), Scottish political economist

Chapter 12: Science on the Fringe of Europe: Eighteenth-Century Sweden

12.1 Carl Linnaeus (1707–1778), Swedish botanist and taxonomist
12.2 C. W. Scheele (1742–1786), Swedish chemist
12.3 E. D. Clarke (1769–1822), English mineralogist and traveller

Chapter 13: Science in Orthodox Europe

13.1 L. Euler (1707–1783), Swiss mathematician, astronomer and physicist
13.2 S. P. Krasheninnikov (1713–1755), Russian natural historian
13.3 Mikhail Lomonosov (1711–1765), Russian natural philosopher

Chapter 14: Establishing Science in Eighteenth-Century Europe

14.1 Robert Jameson (1774–1854), Scottish geologist

Chapter 15: The Chemical Revolution

15.1 Stephen Hales (1677–1761), English chemist and cleric
15.2, 15.3 Joseph Priestley (1733–1804),
English chemist and social reformer
15.4 Antoine Lavoisier (1743–1794), French chemist

Chapter 16: Conclusions

16.1 F. Voltaire (1694–1778), French philosopher
16.2 B. Fontenelle (1657–1757), French philosopher
16.3 Frederick II (1712–1786), Prussian monarch

Acknowledgements

The editor and publishers wish to thank the following for permission to use copyright material:

Cambridge University Press for R. Descartes, *The World* [c. 1629–33], 1664, 1667, *Treatise on Man* [c. 1629–33], 1664, *Discourse on Method*, 1637, *Principles of Philosophy*, 1644, in *The Philosophical Writings of Descartes*, vol. 1, trans. John Cottingham, Robert Stoothoff and Dugald Murdoch (1985), pp. 90–7, 99–108, 118–22, 140, 226–38; Colin Chant for his translation of Mikhail Lomonosov, *Izbrannye priizvedeniya, tom 1: Estesvennye nauki Ifilosofia* [*The Appearance of Venus on the Sun, Observed at the St. Petersburg Academy of Sciences on the 16th Day of May in the Year 1761*] (Moscow: Nauka, 1986), pp. 333–6; Doubleday, a division of Random House, Inc., for material from Galileo Galilei, *Discoveries and Opinions of Galileo*, trans. Stillman Drake, pp. 37–8, 274–7. Copyright © 1957 by Stillman Drake; Dover Publications, Inc., for material from Georgius Agricola, 'On the Knowledge of the Miner', Bk I, *De Re Metallica*, Ulrich Riilein von Calw, 'On the Origin of Metals', in *Ein nützlic Bergbüchlein* [c. 1500], Georgius Agricola, 'On the Origin of Metals', and Georgius Agricola, 'On Assaying', Bk VII, *De Re Metallica*, 1556, in *De Ortu et Causis Subterraneorum*, 1546, trans. H. Clark Hoover and L. Henry Hoover (1912), from Georgius Agricola, *De Re Metallica*, 1556 (Dover reprint, 1950), pp. 1–4, 44–6, 51, 219–20, 222–4; I. Newton, *Opticks*, 1704, 1730 edn (Dover reprint, 1952), pp. 26–33; Antoine Lavoisier, *Elements of Chemistry*, 1789, trans. R. Kerr, 1790 (Dover reprint, 1965), pp. 32–47; and W. Gilbert, *De Magnete*, 1600, trans. P. Fleury Mottelay (Dover reprint, 1958), pp. xxvii–xlv, 23–5, 327, 333–5; Harvard University Press and the Trustees of the Loeb Classical Library for material from Galen, *On the Natural Faculties*, Loeb Classical Library [registered trademark of the President and Fellows of Harvard College], vol. L 71, trans. A. J. Brock (1916), pp. 56–61; the Johns Hopkins University Press for material from Nicholas Copernicus, *On the Revolutions of the Heavenly Spheres*, 1543, in *Copernicus: Complete Works*, vol. 11, ed. Edward Rosen (1978), pp. xix, xx, xxi, 3–6, 7–8; Paracelsus, *Four Treatises of Theophrastus von Hohenheim called*

Paracelsus, ed. Henry Sigerist, 1941 (Princeton University Press, 1996), pp. 16–24, 61–4; and William Edgar Knowles Middleton, *The Experimenters: A Study of the Accademia del Cimento* (1972), pp. 87, 89–92; Manchester University Press for material from *The Leibniz–Clarke Correspondence, 1715–16*, ed. H. G. Alexander (1956), pp. 184–8; Octagon Books, a division of Hippocrene Books, Inc., for material from B. Pascal, *Story of the Great Experiment on the Equilibrium of Fluids*, 1648, trans. I. Spiers and A. Spiers in *The Physical Treatises of Pascal* (Octagon Books, 1973), pp. 97–112; The Open University for material from J. Kepler, *Astronomia Nova*, Heidelberg, 1609, trans. C. A. Russell, in *Science and Religious Belief 1600–1900* (John Wright and Open University Press, 1973), pp. 22–3; and Frederick II, *Oeuvres de Frederic le Grand*, vol. IX, in *The Enlightenment, Texts 1*, trans. S. Eliot and K. Whitlock, A206 (1992), pp. 66–7; Oxford University Press for material from Adam Smith, *Essays on Philosophical Subjects*, ed. W. P. D. Wightman, J. C. Bryce and I. S. Ross. Copyright © 1980 Oxford University Press; Palgrave for material from Tycho Brahe, *De disciplines mathematicis oratio*, 1574, Cornelius Agrippa, *De Occulta Philosophia*, c. 1510, and Oswald Croll, *De signatures internis rerum*, 1608, in *The Occult in Early Modern Europe*, trans. and ed. P. Maxwell-Stuart (1999), pp. 84–5, 96–7, 150–1; Penguin UK for material from Plato, *Timaeus and Critias*, trans. Desmond Lee (Penguin Classics, 1965, rev. 1977). Copyright © H. D. P. Lee 1965, 1971, 1977; Lucretius, *On the Nature of the Universe*, trans. R. E. Latham (Penguin Classics, 1951). Copyright © R. Latham 1951; and with Princeton University Press for Pietro Redondi, *Galileo Heretic*, first published as *Galileo Eretico* by Giulio Einaudi s.p.a. (1983), trans. Raymond Rosenthal (Allen Lane: The Penguin Press, 1988). Italian edition copyright © Giulo Einaudi s.p.a. 1983. This translation copyright © Princeton University Press 1987; Pickering & Chatto Publishers Ltd for material from Robert Boyle, *Some Considerations of the Usefulnesse of Experimental Natural Philosophy, Pt II*, 1671, and Robert Boyle, *Of the Excellency and Grounds of the Corpuscular Philosophy*, 1674, in *The Works of Robert Boyle*, vols. 6 and 8, ed. M. Hunter and E. Davis (2000), pp. 396–400, 103–9; Princeton University Press for material from R. Boyle, 'Dialogue on the Transmutation and Melioration of Metals', *c*. mid- to late 1670s, in L. Principe, *The Aspiring Adept: Robert Boyle and his Alchemical Quest*, pp. 278–88. Copyright © 1998 by Princeton University Press; Aristotle, *Physics* and *On the Heavens* from Aristotle, *Complete Works*, vol. 1, ed. Jonathan Barnes, pp. 342–3, 354–5, 448–51, 458–61, 487–8. Copyright © 1984 by Princeton University Press; Royal Netherlands Academy of Arts and Sciences for material from G. J. Rheticus, *Holy Scripture and the Motion of the Earth*, c. 1540, trans. R. Hooykas, from G. J. Rheticus, *Treatise on Holy Scripture and the Motion of the Earth* (Amsterdam, 1984), pp. 91–101; the Royal Society of Edinburgh for material from James Hutton, *Abstract of a dissertation read in the Royal Society of Edinburgh... concerning the system of the earth, its duration and stability*, repr. in *The 1785 Abstract*, intro. J. Craig (Scottish Academic Press, 1987), pp. 3–30; University of California Press for material from Tommasso Campanella, *Civitas Solis* (City of the Sun), 1623, trans. and intro. Daniel Donno (1981), pp. 27–33, 43–4; Isaac Newton, *The Principia: Mathematical Principles of Natural Philosophy*, 1687, Author's Preface to the Reader, 8 May 1686, trans. and ed. I. B. Cohen and Anne Whitman, pp. 381–3, 391–9. Copyright © 1999 The Regents of the University of California; and Galileo Galilei, *Dialogue Concerning the Two Chief World Systems: The Ptolemaic and Copernican*, 1632, trans. Stillman Drake, pp. 138–49. Copyright © 1952, 1962, 1967 The Regents of the University of California; the University of Chicago Press for material from Sacrobosco, *The Sphere*, in *The Sphere of Sacrobosco and its Commentators*, trans. Lynn Thorndike (1949), pp. 118–22; the University of Wisconsin Press for material from *The Correspondence of Henry Oldenburg, vol. III (1666–1667)*, trans. and ed. A. Rupert Hall and Marie Boas Hall (1966), pp. 141–2.

Every effort has been made to trace the copyright holders but if any have been inadvertently overlooked the publishers will be pleased to make the necessary arrangement at the first opportunity.

Preface

This Reader provides a documentary history of what is still widely regarded as perhaps the most significant event in human history since the rise of Christianity, namely the Scientific Revolution. The Reader attempts to explore through primary sources both the conceptual and the institutional foundations of modern science that were themselves built upon the scientific and natural philosophical heritage of classical antiquity. Insofar as an international scientific community was beginning to emerge by 1800, the nature of early modern European societies and states in the preceding three centuries remains a crucial arena for historical investigation. It has been increasingly evident over the last two decades that an ever richer social, cultural and intellectual contextualisation of national and cultural traditions has both sharpened and reshaped our understanding of what was clearly an uneven spread of science across the continent. The distinctive character of national and local contexts has been a signal feature of the Open University undergraduate course and accompanying textbook of the same title, *The Rise of Scientific Europe 1500–1800*, ed. David Goodman and Colin A. Russell (Hodder & Stoughton/The Open University, 1991). The emphasis on a cultural topography that explores the interplay of social, political, religious and scientific ideas underpins the notion, reflected here, of distinctive national roles and styles of science. Though this Reader is designed to support students following the course in its newly expanded form, the prominence of key themes embracing ancients and moderns, science and religion, scholars and craftsmen, scientific patronage, the occult sciences, matter and motion, and historiography suggests that this compilation should be of interest to a wider undergraduate and general readership.

The present volume incorporates some materials included in an Anthology compiled and edited for the predecessor course by Colin Russell, which was not generally available outside of The Open University. This new Reader covers a considerably more extensive range of readings relating to the key themes already mentioned and signposts both familiar and less familiar texts associated with the rise of new approaches to the physical world across Europe. Virtually all the sciences are touched upon in this Reader at some point, from the mathematically based classical fields of astronomy, optics and mechanics to the more applied sciences, such as chemistry, metallurgy, experimental physics and geology. The rather sparser coverage of medicine, natural history and botany reflects the basic make-up of the original course rather than any predisposition to marginalise those very important areas. In practice, the boundaries between these categories were constantly renegotiated, as were the spaces between practical application and theoretical transformation. Similarly, frequent and complex associations with the occult sciences remained fluid and open for much of this period. The Scientific Revolution has been characterised as a process of change and displacement within competing systems of natural philosophy. In general terms, the shape, direction and legitimacy of the sciences were viewed by early modern contemporaries as part of a larger system of one or more of the natural philosophies on offer.

While the Reader can be used on its own, it has been designed to be read alongside a companion volume, *Science in Europe, 1500–1800: A Secondary Sources Reader*, ed. Malcolm Oster (Palgrave/The Open University, 2002). The two volumes together provide both cross-fertilisation between primary and secondary sources and a resource for class use. An ever-present danger in readings documenting the period is that they will be judged by a preoccupation as to what constitutes progress and rationality in modern science today. While it would be unrealistic to expect that the historiographical stance of 'whiggism' can ever be fully excised from our current perspectives, the expectation of what early modern thinkers and practitioners themselves believed counted as knowledge should act as the more desirable criterion in understanding supposed gaps between ideal and achievement.

The readings are grouped under the chapter headings found in the Goodman and Russell textbook, and an attempt has been made to provide a balance of short, medium and long extracts. Footnotes have been retained where possible in contemporary sources, while editorial footnotes in modern editions have been excised or occasionally retained or re-edited for clear sequencing in the extract selected. Original page numbers or ranges of page numbers used are cited in source references, unless chapters are more appropriate. The 'normal' ellipsis of just three dots... indicates a short omission, and a bracketed ellipsis [...] shows that larger chunks of material (at least a paragraph) have been left out. The dates in square brackets on the Contents pages and in source references are dates of *composition*; all other dates are dates of *publication*, except dated letters.

I owe a considerable debt to Colin Russell, Noel Coley, Michael Bartholomew and Peter Morris, who were involved in the original writing of the course and in compiling documents for the accompanying Reader. I also need to register equal thanks to current departmental colleagues, who were similarly involved in the original writing of the course and compiled items for the accompanying Reader: David Goodman, who was responsible for the final selection in Chapters 1, 4, 5, 7 and 12, Colin Chant for Chapter 13 and Gerrylynn Roberts for Chapter 14, in this new Reader. Similarly, I have to thank Kate Crawley and Michael Honeybone for their patience and hard work in providing invaluable feedback on the selection as experienced tutors of the course, and the external assessor, Stephen Pumfrey, in challenging us to meet our learning objectives. Finally, I have to register my debt to the considerable assistance of Gill Gowans, Alison Kirkbright and Shirley Coulson in preparing the typescript of this volume.

Malcolm Oster

Chapter One
Europe's Awakening

1.1 Plato, *Timaeus*, trans. H. D. P. Lee (Harmondsworth: Penguin, 1965), pp. 40–5, 71–8

[...]

TIMAEUS: Yes, Socrates; of course everyone with the least sense always calls on god at the beginning of any undertaking, small or great. So surely, if we are not quite crazy, as we embark on our account of how the universe began, or perhaps had no beginning, we must pray to all the gods and goddesses that what we say will be pleasing to them first, and then to ourselves. Let that be our invocation to the gods: but we must invoke our own powers too, that you may most easily understand and I most clearly expound my thoughts on the subject before us.

We must in my opinion begin by distinguishing between that which always is and never becomes from that which is always becoming but never is. The one is apprehensible by intelligence with the aid of reasoning, being eternally the same, the other is the object of opinion and irrational sensation, coming to be and ceasing to be, but never fully real. In addition, everything that becomes or changes must do so owing to some cause; for nothing can come to be without a cause. Whenever, therefore, the maker of anything keeps his eye on the eternally unchanging and uses it as his pattern for the form and function of his product the result must be good; whenever he looks to something that has come to be and uses a model that has come to be, the result is not good.

As for the world – call it that or 'cosmos' or any other name acceptable to it – we must ask about it the question one is bound to ask to begin with about anything: whether it has always existed and had no beginning, or whether it has come into existence and started from some beginning. The answer is that it has come into being; for it is visible, tangible, and corporeal, and therefore perceptible by the senses, and... sensible things are objects of opinion and sensation and therefore change and come into being. And what comes into being or changes must do so, we said, owing to some cause. To discover the maker and father of this universe is indeed a hard task, and having found him it would be impossible to tell everyone about him. Let us return to our question, and ask to which pattern did its constructor work, that which remains the same and unchanging or that which has come to be? If the world is beautiful and its maker good, clearly he had his eye on the eternal; if the alternative (which it is blasphemy even to mention) is true, on that which is subject to change. Clearly, of course, he had his eye on the eternal; for the world is the fairest of all things that have come into being and he is the best of causes. That being so, it must have been constructed on the pattern of what is apprehensible by reason and understanding and eternally unchanging; from which again it follows that the world is a likeness of something else.... Don't... be surprised, Socrates, if on many matters concerning the gods and the whole world of change we are unable in every respect and on every occasion to render a consistent and accurate account. You must be satisfied if our account is as likely as any, remembering that both I and you who are sitting in judgement on it are merely human, and should not look for anything more than a likely story in such matters.

SOCRATES: Certainly, Timaeus; we must accept your principles in full. You have given us a wonderfully acceptable prelude; now go on to develop your main theme.

TIMAEUS: Let us therefore state the reason why the framer of this universe of change framed it at all. He was good, and what is good has no particle of envy in it; being therefore without envy he wished all things to be as like himself as possible. This is as valid a principle for the origin of the world of change as we shall discover from the wisdom of men, and we should accept it. God therefore, wishing that all things should be good, and so far as possible nothing be imperfect, and finding the visible universe in a state not of rest but of inharmonious and disorderly motion, reduced it to order from disorder, as he judged that order was in every way better. It is impossible for the best to produce anything but the highest. When he considered, therefore, that in all the realm of visible nature, taking each thing as a whole, nothing without intelligence is to be found that is superior to anything with it, and that intelligence is impossible without soul, in fashioning the universe he implanted reason in soul and soul in body, and so ensured that his work should be by nature highest and best. And so that most likely account must say that this world came to be in very truth, through god's providence, a living being with soul and intelligence.

On this basis we must proceed to the next question: What was the living being in the likeness of which the creator constructed it? We cannot suppose that it was any creature that is part of a larger whole, for nothing can be good that is modelled on something incomplete.... For god's purpose was to use as his model the highest and most completely perfect of intelligible things, and so he created a single visible living being, containing within itself all living beings of the same natural order. Are we then right to speak of one universe, or would it be more correct to speak of a plurality or infinity? ONE is right, if it was manufactured according to its pattern; for that which comprises all intelligible beings cannot have a double. There would have to be another being comprising them both, of which both were parts, and it would be correct to call our world a copy not of them but of the being which comprised them. In order therefore that our universe should resemble the perfect living creature in being unique, the maker did not make two universes or an infinite number, but our universe was and is and will continue to be his only creation.

Now anything that has come to be must be corporeal, visible, and tangible: but nothing can be visible without fire, nor tangible without solidity, and nothing can be solid without earth. So god, when he began to put together the body of the universe, made it of fire and earth. But it is not possible to combine two things properly without a third to act as a bond to hold them together. And the best bond is one that effects the closest unity between itself and the terms it is combining; and this is best done by a continued geometrical proportion.... If then the body of the universe were required to be a plane surface with no depth, one middle term would have been enough to connect it with the other terms, but in fact it needs to be solid, and solids always need two connecting middle terms. So god placed water and air between fire and earth, and made them so far as possible proportional to one another, so that air is to water as water is to earth; and in this way he bound the world into a visible and tangible whole. So by these means and from these four constituents the body of the universe was created to be at unity owing to proportion; in consequence it acquired concord, so that having once come together in unity with itself it is indissoluble by any but its compounder.

The construction of the world used up the whole of each of these four elements. For the creator constructed it of all the fire and water and air and earth available, leaving over no part or property of any of them, his purpose being, firstly, that it should be as complete a living being as possible, a whole of complete parts, and further, that it should be single and there should be nothing left over out of which another such whole could come into being, and finally that it should be ageless and free from disease. For he knew that heat and cold and other things that have powerful effects attack a composite body from without, so causing untimely dissolution, and make it decay by bringing disease and old age upon it. On this account and for this reason he made this world a single complete whole, consisting of parts that are wholes, and subject neither to age nor to disease. The shape he gave it was suitable to its nature. A suitable shape for a living being that was to contain within itself all living beings would be a figure that contains all possible figures within itself. Therefore he turned it into a rounded spherical shape, with the extremes equidistant in all directions from the centre, a figure that has the greatest degree of completeness and uniformity, as he judged uniformity to be incalculably superior to its opposite. And he gave it a perfectly smooth external finish all round, for many reasons. For it had no need of eyes, as there remained nothing visible outside it, nor of hearing, as there remained nothing audible; there was no surrounding air which it needed to breathe in, nor was it in need of any organ by which to take food into itself and discharge it later after digestion. Nothing was taken from it or added to it, for there was nothing that could be; for it was designed to supply its own nourishment from its own decay and to comprise and cause all processes, as its creator thought that it was better for it to be self-sufficient than dependent on anything else. He did not think there was any purpose in providing it with hands as it had no need to grasp anything or defend itself, nor with feet or any other means of support. For of the seven[1] physical motions he allotted to it the one which most properly belongs to intelligence and reason, and made it move with a uniform circular motion on the same spot; any deviation into movement of the other six kinds he entirely precluded. And because for its revolution it needed no feet he created it without feet or legs.

This was the plan of the eternal god when he gave to the god about to come into existence a smooth and unbroken surface, equidistant in every direction from the centre, and made it a physical body whole and complete, whose components were also complete physical bodies. And he put soul in the centre and diffused it through the whole and enclosed the body in it. So he established a single spherical universe in circular motion, alone but because of its excellence needing no company other than itself, and satisfied to be its own acquaintance and friend. His creation, then, for all these reasons, was a blessed god. [...]

My verdict, in short, may be stated as follows. There were, before the world came into existence, being, space, and becoming, three distinct realities. ...And its contents were in constant process of movement and separation, rather like the contents of a winnowing basket or similar implement for cleaning corn, in which the solid and heavy stuff is sifted out and settles on one side, the light and insubstantial on another: so the four basic constituents were shaken by the receptacle, which acted as a kind of shaking implement, and those most unlike each other were separated most widely, those most like each other pushed together most closely, with the result that they came to occupy different regions of space even before they were arranged into an ordered universe. Before that time they were all without proportion or measure; fire, water, earth and air bore some traces of their proper nature,

1 The seven motions are: uniform circular motion in the same place, mentioned here, up and down, backwards and forwards, right and left.

but were in the disorganized state to be expected of anything which god has not touched, and his first step when he set about reducing them to order was to give them a definite pattern of shape and number. We must thus assume as a principle in all we say that god brought them to a state of the greatest possible perfection, in which they were not before. Our immediate task is to attempt an explanation of the particular structure and origin of each; its terms will be unfamiliar, but you will be able to follow as you have been trained in the branches of knowledge which it must employ.

In the first place it is clear to everyone that fire, earth, water, and air are bodies, and all bodies are solids. All solids again are bounded by surfaces, and all rectilinear surfaces are composed of triangles. There are two basic types of triangle, each having one right angle and two acute angles: in one of them these two angles are both half right angles, being subtended by equal sides, in the other they are unequal, being subtended by unequal sides. This we postulate as the origin of fire and the other bodies, our argument combining likelihood and necessity; their more ultimate origins are known to god and to men whom god loves. We must proceed to enquire what are the four most perfect possible bodies which, though unlike one another, are some of them capable of transformation into each other on resolution. If we can find the answer to this question we have the truth about the origin of earth and fire and the two mean terms between them; for we will never admit that there are more perfect visible bodies than these, each in its type. So we must do our best to construct four types of perfect body and maintain that we have grasped their nature sufficiently for our purpose. Of the two basic triangles, then, the isosceles has only one variety, the scalene an infinite number.

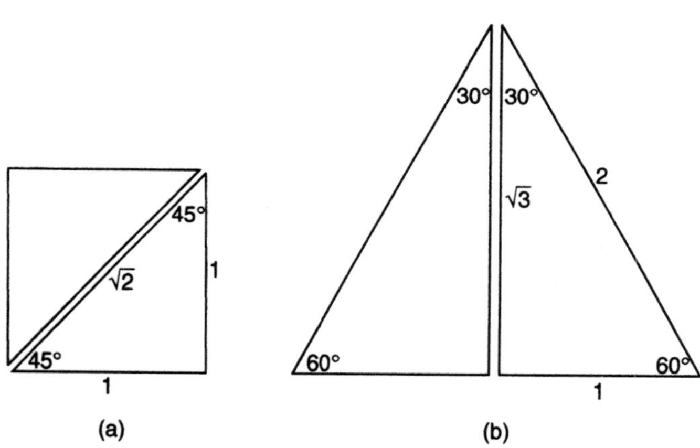

The two types of triangle: (a) isosceles = a triangle with two sides of equal length; (b) scalene = a triangle with all three sides of different lengths.

We must therefore choose, if we are to start according to our own principles, the most perfect of this infinite number. If anyone can tell us of a better choice of triangle for the construction of the four bodies, his criticism will be welcome; but for our part we propose to pass over all the rest and pick on a single type, that of which a pair compose an equilateral triangle. It would be too long a story to give the reason....So let us assume that these are the two triangles from which fire and the other bodies are constructed, one isosceles and the other having a greater side whose square is three times that of the lesser. We must now proceed to clarify something we left undetermined a moment ago. It appeared as if all four types of body could pass into each other in the process of change; but this appearance is misleading. For, of the four bodies that are produced by our chosen types of triangle, three are composed of the scalene, but the fourth alone from the isosceles. Hence all four cannot pass into each other...; this can only happen with three of them. For these are all composed of one triangle, and when larger bodies are broken up a number of small bodies are formed of the same constituents, taking on their appropriate figures; and when small bodies are broken up into their component triangles a single new larger figure may be formed as they are unified into a single solid.[1]

So much for their transformation into each other. We must next describe what geometrical figure each body has and what is the number of its components. We will begin with the construction of the simplest and smallest figure. Its basic unit is the triangle whose hypotenuse is twice the length of its shorter side. If two of these are put together with the hypotenuse as diameter of the resulting figure, and if the process is repeated three times and the diameters and shorter sides of the three figures are made to coincide in the same vertex, the result is a single equilateral triangle composed of six basic units. And if four equilateral triangles are put together, three of their plane angles meet to form a single solid angle....

The second figure is composed of the same basic triangles put together to form eight equilateral triangles, which yield a single solid angle from four planes. The formation of six such solid angles completes the second figure.

The third figure is put together from one hundred and twenty basic triangles, and has twelve solid angles, each bounded by five equilateral plane triangles, and twenty faces, each of which is an equilateral triangle.

After the production of these three figures the first of our basic units is dispensed with, and the isosceles triangle is used to produce the fourth body. Four such triangles are put together with their right angles meeting at

---

1 The three sentences are very compressed and to some extent anticipate what we shall shortly be told about the distribution of regular solids between the elements. The process of transformation is thought of as the breaking down of a regular solid into its constituent triangles, which can then rejoin to form a solid of different figure. From this process the cube (earth) must be excluded as its constituent triangle is of a different type to that of the other three.

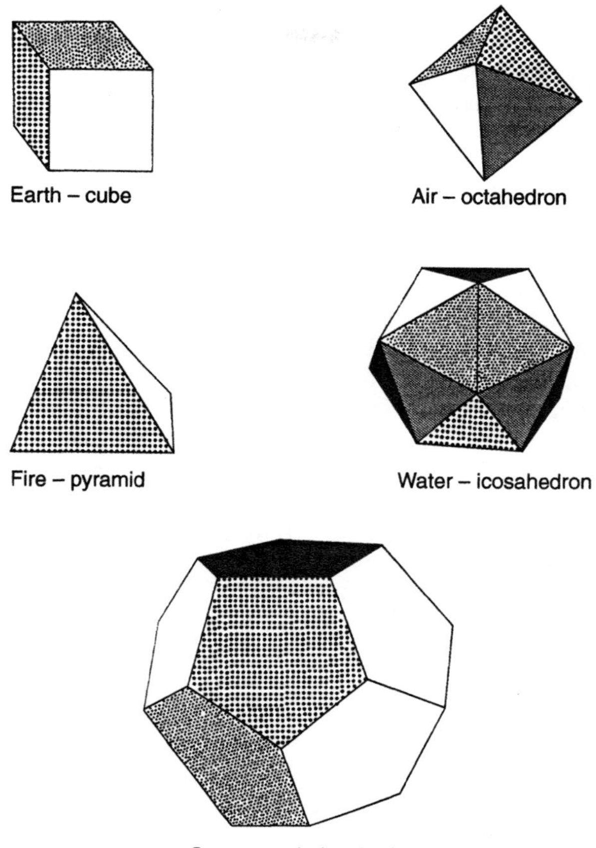

The four figures are: the pyramid, the octahedron, the icosahedron, and the cube; the fifth [is] the dodecahedron. The dodecahedron cannot be constructed out of the basic triangles and, because it approaches the sphere most nearly in volume, is associated here with 'the whole (spherical) heaven', just as it is associated in the Phaedo . . . *with the spherical earth. Exactly how Plato supposed god used it for 'arranging the constellations' (literally 'embroidering with figures') we are not told.*

a common vertex to form a square. Six squares fitted together complete eight solid angles, each composed by three plane right angles. The figure of the resulting body is the cube, having six plane square faces.

There still remained a fifth construction, which the god used for arranging the constellations on the whole heaven.[. . .]

We must proceed to distribute the figures whose origins we have just described between fire, earth, water, and air. Let us assign the cube to earth; for it is the most immobile of the four bodies and the most retentive of shape, and these are characteristics that must belong to the figure with the most stable faces. And of the basic triangles we have assumed, the isosceles has a naturally more stable base than the scalene, and of the equilateral figures composed of them the square is, in whole and in part, a firmer base than the equilateral triangle. So we maintain our principle of likelihood by assigning it to earth, while similarly we assign the least mobile of the other figures to water, the most mobile to fire, and the intermediate to air. And again we assign the smallest figure to fire, the largest to water, the intermediate to air; the sharpest to fire, the next sharpest to air, and the least sharp to water. So to sum up, the figure which has the fewest faces must in the nature of things be the most mobile, as well as the sharpest and most penetrating, and finally, being composed of the smallest number of similar parts, the lightest. Our second figure will be second in all these respects, our third will be third. Logic and likelihood thus both require us to regard the pyramid as the solid figure that is the basic unit or seed of fire; and we may regard the second of the figures we constructed as the basic unit of air, the third of water. We must, of course, think of the individual units of all four bodies as being far too small to be visible, and only becoming visible when massed together in large numbers; and we must assume that the god duly adjusted the proportions between their numbers, their movements, and their other qualities and brought them in every way to the exactest perfection....

1.2 Aristotle, (a) *Physics*, (b) *On the Heavens*, from Aristotle, *Complete Works*, vol. 1, ed. Jonathan Barnes (Princeton, NJ: Princeton University Press, 1984), (a) pp. 342–3, 354–5, (b) pp. 448–51, 458–61, 487–8

(a) Physics

BOOK III

Nature is a principle of motion and change, and it is the subject of our inquiry. We must therefore see that we understand what motion is; for if it were unknown, nature too would be unknown.

When we have determined the nature of motion, our task will be to attack in the same way the terms which come next in order. Now motion is supposed to belong to the class of things which are continuous; and... place, void, and time are thought to be necessary conditions of motion....

To begin then, as we said, with motion.

Some things are in fulfilment only, others in potentiality and in fulfilment....

We have distinguished... between what is in fulfilment and what is potentially; thus the fulfilment of what is potentially, as such, is motion – e.g. the fulfilment of what is alterable, as alterable, is alteration; of what is increasable and its opposite, decreasable . . . , increase and decrease; of what can come to be and pass away, coming to be and passing away; of what can be carried along, locomotion.

That this is what motion is, is clear from what follows: when what is buildable, in so far as we call it such, is in fulfilment, it is being built, and that is building. Similarly with learning, doctoring, rolling, jumping, ripening, aging.

The same thing can be both potential and fulfilled, not indeed at the same time or not in the same respect, but e.g. potentially hot and actually cold. Hence such things will act and be acted on by one another in many ways: each of them will be capable at the same time of acting and of being acted upon. Hence, too, what effects motion as a natural agent can be moved: when a thing of this kind causes motion, it is itself also moved. This, indeed, has led some people to suppose that every mover is moved. . . . It *is* possible for a thing to cause motion, though it is itself incapable of being moved.

It is the fulfilment of what is potential when it is already fulfilled and operates not as itself but as movable, that is motion. What I mean by 'as' is this: bronze is potentially a statue. But it is not the fulfilment of bronze as *bronze* which is motion. . . .

It is evident that this is motion, and that motion occurs just when the fulfilment itself occurs, and neither before nor after. For each thing is capable of being at one time actual, at another not. Take for instance the buildable: the actuality of the buildable as buildable is the process of building. For the actuality must be either this or the house. But when there is a house, the buildable is no longer there. On the other hand, it *is* the buildable which is *being* built. Necessarily, then, the actuality is the process of building. But building is a kind of motion, and the same account will apply to the other kinds also. [. . .]

BOOK IV

The physicist must have a knowledge of place, too, as well as of the infinite – namely, whether there is such a thing or not, and the manner of its existence and what it is – both because all suppose that things which exist are *somewhere* (the non-existent is nowhere – where is the goat-stag or the sphinx?), and because motion in its most general and proper sense is change of place, which we call 'locomotion'.

The question, what is place? presents many difficulties. An examination of all the relevant facts seems to lead to different conclusions. . . .

The existence of place is held to be obvious from the fact of mutual replacement. Where water now is, there in turn, when the water has gone out as from a vessel, air is present; and at another time another body occupies this same place. The place is thought to be different from all the bodies which come to be in it and replace one another. What now contains air formerly contained water, so that clearly the place or space into which and out of which they passed was something different from both.

Further, the locomotions of the elementary natural bodies – namely, fire, earth, and the like – show not only that place is something, but also that it exerts a certain influence. Each is carried to its own place, if it is not hindered, the one up, the other down. Now these are regions or kinds of place – up and down and the rest of the six directions. Nor do such distinctions (up and down and right and left) hold only in relation to us. To *us* they are not always the same but change with the direction in which we are turned: that is why the same thing is often both right *and* left, up *and* down, before *and* behind. But in *nature* each is distinct, taken apart by itself. It is not every chance direction which is up, but where fire and what is light are carried; similarly, too, down is not any chance direction but where what has weight and what is made of earth are carried – the implication being that these places do not differ merely in position, but also as possessing distinct powers....

These considerations then would lead us to suppose that place is something distinct from bodies, and that every sensible body is in place.... If this is its nature, the power of place must be a marvellous thing, and be prior to all other things. For that without which nothing else can exist, while it can exist without the others, must needs be first; for place does not pass out of existence when the things in it are annihilated. [...]

(b) On the Heavens

[...] All natural bodies and magnitudes we hold to be, as such, capable of locomotion; for nature, we say, is their principle of movement. But all movement that is in place, all locomotion, as we term it, is either straight or circular or a combination of these two which are the only simple movements. And the reason is that these two, the straight and the circular line, are the only simple magnitudes. Now revolution about the centre is circular motion, while the upward and downward movements are in a straight line, 'upward' meaning motion away from the centre, and 'downward' motion towards it. All simple motion, then, must be motion either away from or towards or about the centre....

Bodies are either simple or compounded of such; and by simple bodies I mean those which possess a principle of movement in their own nature, such as fire and earth with their kinds, and whatever is akin to them. Necessarily, then, movements also will be either simple or in some sort compound – simple in the case of the simple bodies, compound in that of the composite – and the motion is according to the prevailing element. Supposing, then, that there is such a thing as simple movement, and that circular movement is simple... then there must necessarily be some simple body which moves naturally and in virtue of its own nature with a circular movement. ...Further, this circular motion is necessarily primary. For the complete is naturally prior to the incomplete, and the circle is a complete thing. This cannot be said of any straight line: – not of an infinite line; for then it would have a limit and an end: nor of any finite line; for in every case there is something beyond it, since any finite line can be extended. And so, since the prior movement belongs to the body which is naturally prior, and circular movement is prior to straight, and movement in a straight line belongs to simple bodies – fire moving straight upward and earthy bodies straight downward towards the centre – since this is so, it follows that circular movement also must be the movement of some simple body. For the movement of composite bodies is, as we said, determined by that simple body which prevails in the composition. From this it is clear that there is in nature some bodily substance other than the formations we know, prior to them all and more divine than they.... Further, if, on the one hand, circular movement is *natural* to something, it must surely be some simple and primary body which naturally moves with a natural circular motion, as fire moves up and earth down. If, on the other hand, the movement of the rotating bodies about the centre is *unnatural*, it would be remarkable and indeed quite inconceivable that this movement alone should be continuous and eternal, given that it is unnatural. At any rate the evidence of all other cases goes to show that it is the unnatural which quickest passes away.... On all these grounds, therefore, we may infer with confidence that there is something beyond the bodies that are about us on this earth, different and separate from them; and that the superior glory of its nature is proportionate to its distance from this world of ours.

In consequence of what has been said,... it is clear that not every body possesses either lightness or heaviness. We must explain in what sense we are using the words 'heavy' and 'light'.... Let us then apply the term 'heavy' to that which naturally moves towards the centre, and 'light' to that which moves naturally away from the centre. The heaviest thing will be that which sinks to the bottom of all things that move downward, and the lightest that which rises to the surface of everything that moves upward. Now, necessarily, everything which moves either up or down possesses lightness or heaviness or both – but not both relatively to the same thing; for things are heavy and light relatively to one another; air, for instance, is light relatively to water, and water light relatively to earth. But the body which moves in a circle cannot possibly possess heaviness or lightness. For neither naturally nor unnaturally can it move either towards or away from the centre. Movement in a straight line certainly does not belong to it *naturally*, since one sort of movement is, as we saw, appropriate to each simple body, and so we should be compelled to identify it with one of the bodies which move in this way....

It is equally reasonable to assume that this body will be ungenerated and indestructible and exempt from increase and alteration.... Now the motions of contraries are contrary. If then this body can have no contrary, because there can be no contrary motion to the circular, nature seems justly to have exempted from contraries the body which was to be ungenerated and indestructible. For it is on contraries that generation and destruction depend. Again, that which is subject to increase increases upon contact with a kindred body, which is resolved into its matter. But there is nothing out of which this body can have been generated. And if it is exempt from increase and destruction, the same reasoning leads us to suppose that it is also unalterable. For alteration is movement in respect of quality; and qualitative states and dispositions, such as health and disease, do not come into being without changes of properties. But all natural bodies which change their properties we see to be subject to increase and diminution. This is the case, for instance, with the bodies of animals and their parts and with vegetable bodies, and similarly also with those of the elements. And so, if the body which moves with a circular motion cannot admit of increase or diminution, it is reasonable to suppose that it is also unalterable.

The reasons why the primary body is eternal and not subject to increase or diminution, but unaging and unalterable and unmodified, will be clear from what has been said to any one who believes in our assumptions. Our theory seems to confirm the phenomena and to be confirmed by them. For all men have some conception of the nature of the gods, and all who believe in the existence of gods at all, whether barbarian or Greek, agree in allotting the highest place to the deity, surely because they suppose that immortal is linked with immortal and regard any other supposition as impossible. If then there is, as there certainly is, anything divine, what we have just said about the primary bodily substance was well said. The mere evidence of the senses is enough to convince us of this, at least with human certainty. For in the whole range of time past, so far as our inherited records reach, no change appears to have taken place either in the whole scheme of the outermost heaven or in any of its proper parts. The name, too, of that body seems to have been handed down right to our own day from our distant ancestors who conceived of it in the fashion which we have been expressing. The same ideas, one must believe, recur in men's minds not once or twice but again and again. And so, implying that the primary body is something else beyond earth, fire, air, and water, they gave the highest place the name of *aether*, derived from the fact that it 'runs always' for an eternity of time....

It is also clear from what has been said why the number of what we call simple bodies cannot be greater than it is. The motion of a simple body must itself be simple, and we assert that there are only these two simple motions, the circular and the straight, the latter being subdivided into motion away from and motion towards the centre.

[...] [W]hatever possesses weight or lightness will have its place either at one of the extremes or in the middle region. But this is impossible while the world is conceived as infinite. And, generally, that which has no centre or extreme limit, no up or down, gives the bodies no place for their motion; and without that movement is impossible. A thing must move either naturally or unnaturally, and the two movements are determined by the proper and alien places. Again, a place in which a thing rests or to which it moves unnaturally, must be the natural place for some other body.... From these arguments then it is clear that the body of the universe is not infinite.

We must now proceed to explain why there cannot be more than one heaven....

... [T]he elements must also be the same everywhere. The particles of earth, then, in another world move naturally also to our centre and its fire to our circumference. This, however, is impossible, since, if it were true, earth must, in its own world, move upwards, and fire to the centre; in the same way the earth of our world must move naturally away from the centre when it moves towards the centre of another universe. This follows from the supposed juxtaposition of the worlds. For either we must refuse to admit the identical nature of the simple bodies in the various universes, or, admitting this, we must make the centre and the extremity one as suggested. This being so, it follows that there cannot be more worlds than one. [...]

A consideration of the other kinds of movement also makes it plain that there is some point to which earth and fire move naturally. For in general that which is moved changes from something into something, the starting-point and the goal being different in form, and always it is a finite change. For instance, to recover health is to change from disease to health, to increase is to change from smallness to greatness. Locomotion must be similar; for it also has its goal and starting-point – and therefore the starting-point and the goal of the natural movement must differ in form – just as the movement of coming to health does not take any direction which chance or the wishes of the mover may select. Thus, too, fire and earth move not to infinity but to opposite points; and since the opposition in place is between above and below, these will be the limits of their movement.... There must therefore be some end to locomotion: it cannot continue to infinity.

This conclusion that local movement is not continued to infinity is corroborated by the fact that earth moves more quickly the nearer it is to the centre, and fire the nearer it is to the upper place. But if movement were infinite, speed would be infinite also; and if speed then weight and lightness. [...]

We must show not only that the heaven is one, but also that more than one heaven is impossible, and, further, that, as exempt from decay and generation, the heaven is eternal. [...]

Let us first decide the question whether the earth moves or is at rest. For, as we said, there are some who make it one of the stars, and others who, setting it at the centre, suppose it to be rolled and in motion about the pole as axis. That both views are untenable will be clear if we take as our starting-point the fact that the earth's motion, whether the earth be at the centre or away from it, must needs be a constrained motion. It cannot be the movement of the earth itself. If it were, any portion of it would have this movement; but in fact every part moves in a straight line to the centre. Being, then, constrained and unnatural, the movement could not be eternal. But the order of the universe is eternal. Again, everything that moves with the circular movement, except the first sphere, is observed to be passed, and to move with more than one motion. The earth, then, also, whether it moves about the centre or is stationary at it, must necessarily move with two motions. But if this were so, there would have to be passings and turnings of the fixed stars. Yet no such thing is observed. The same stars always rise and set in the same parts of the earth.

Further, the natural movement of the earth, part and whole alike, is to the centre of the whole – whence the fact that it is now actually situated at the centre – but it might be questioned, since both centres are the same, which centre it is that portions of earth and other heavy things move to. Is this their goal because it is the centre of the earth or because it is the centre of the whole? The goal, surely, must be the centre of the whole. For fire and other light things move to the extremity of the area which contains the centre. It happens, however, that the centre of the earth and of the whole is the same. Thus they do move to the centre of the earth, but accidentally, in virtue of the fact that the earth's centre lies at the centre of the whole.... It is clear, then, that the earth must be at the centre and immovable, not only for the reasons already given, but also because heavy bodies forcibly thrown quite straight upward return to the point from which they started, even if they are thrown to an unlimited distance. From these considerations then it is clear that the earth does not move and does not lie elsewhere than at the centre.

From what we have said the explanation of the earth's immobility is also apparent. If it is the nature of earth, as observation shows, to move from any point to the centre, as of fire contrariwise to move from the centre to the extremity, it is impossible that any portion of earth should move away from the centre except by constraint. For a single thing has a single movement, and a simple thing a simple: contrary movements cannot belong to the same thing, and movement away from the centre is the contrary of movement to it. If then no portion of earth can move away from the centre, obviously still less can the earth as a whole so move. For it is the nature of the whole to move to the point to which the part naturally moves. Since, then, it would require a force greater than itself to move it, it must needs stay at the centre. This view is further supported by the contributions of mathematicians to astronomy, since the phenomena – the changes of the shapes by which the order of the stars is determined – are fully accounted for on the hypothesis that the earth lies at the centre. Of the position of the earth and of the manner of its rest or movement, our discussion may here end.

Its shape must necessarily be spherical. For every portion of earth has weight until it reaches the centre, and the jostling of parts greater and smaller would bring about not a waved surface, but rather compression and convergence of part and part until the centre is reached. The process should be conceived by supposing the earth to come into being in the way that some of the natural philosophers describe.... If, on the one hand, there were a similar movement from each quarter of the extremity to the single centre, it is obvious that the resulting mass would be similar on every side. For if an equal amount is added on every side the extremity of the mass will be everywhere equidistant from its centre, i.e. the figure will be spherical. But neither will it in any way affect the argument if there is not a similar accession of concurrent fragments from every side. For the greater quantity, finding a lesser in front of it, must necessarily drive it on, both having an impulse whose goal is the centre, and the greater weight driving the lesser forward till this goal is reached....

If the earth was generated, then, it must have been formed in this way, and so clearly its generation was spherical....

1.3 Lucretius, *On the Nature of the Universe*, trans. R. E. Latham (Harmondsworth: Penguin, 1951), pp. 31, 33–7, 39, 44–5, 54–7, 63–4, 66, 70, 72–3, 80, 91–2, 177, 195–7

[...] [O]ur starting-point will be this principle: *Nothing can ever be created by divine power out of nothing.* [...]

The second great principle is this: *nature resolves everything into its component atoms and never reduces anything to nothing.* If anything were perishable in all its parts, anything might perish all of a sudden and vanish from sight. There would be no need of any force to separate its parts and loosen their links. In actual fact, since everything is composed of indestructible seeds, nature obviously does not allow anything to perish till it has encountered a force that shatters it with a blow or creeps into chinks and unknits it. [...]

Again, all objects would regularly be destroyed by the same force and the same cause, were it not that they are sustained by imperishable matter more or less tightly fastened together. Why, a mere touch would be enough to bring about destruction supposing there were no imperishable bodies whose union could be dissolved only by the appropriate force. Actually, because the fastenings of the atoms are of various kinds while their matter is imperishable, compound objects remain intact until one of them encounters a force that proves strong enough to break up its particular constitution. Therefore nothing returns to nothing, but everything is resolved into its constituent bodies. [...]

Well, Memmius, I have taught you that things cannot be created out of nothing nor, once born, be summoned back to nothing. Perhaps, however, you are becoming mistrustful of my words, because these atoms of mine are not visible to the eye. Consider, therefore, this further evidence of *bodies whose existence you must acknowledge though they cannot be seen....*

... [W]e smell the various scents of things though we never see them approaching our nostrils. Similarly, heat and cold cannot be detected by our eyes, and we do not see sounds. Yet all these must be composed of bodies, since they are able to impinge upon our senses. For nothing can touch or be touched except body. [...]

Again, in the course of many annual revolutions of the sun a ring is worn thin next to the finger with continual rubbing. Dripping water hollows a stone. A curved ploughshare, iron though it is, dwindles imperceptibly in the furrow. We see the cobble-stones of the highway worn by the feet of many wayfarers. The bronze statues by the city gates show their right hands worn thin by the touch of travellers who have greeted them in passing. We see that all these are being diminished, since they are worn away. But to perceive what particles drop off at any particular time is a power grudged to us by our ungenerous sense of sight. [...]

On the other hand, things are not hemmed in by the pressure of solid bodies in a tight mass. This is because *there is vacuity in things*.... Well then, by vacuity I mean intangible and empty space. If it did not exist, things could not move at all. [...]

To pick up the thread of my discourse, all nature as it is in itself consists of two things – bodies and the vacant space in which the bodies are situated and through which they move in different directions. [...]

To proceed with our argument, there is an ultimate point in visible objects which represents the smallest thing that can be seen. So also there must be an ultimate point in objects that lie below the limit of perception by our senses. This point is without parts and is the smallest thing that can exist. It never has been and never will be able to exist by itself, but only as one primary part of something else. It is with a mass of such parts, solidly jammed together in order, that matter is filled up. Since they cannot exist by themselves, they must needs stick together in a mass from which they cannot by any means be prized loose. The atoms therefore are absolutely solid and unalloyed, consisting of a mass of least parts tightly packed together. They are not compounds formed by the coalescence of their parts, but bodies of absolute and everlasting solidity. To these nature allows no loss or diminution. [...]

And now pay special attention to what follows and listen more intently. ... My art is not without a purpose. Physicians, when they wish to treat children with a nasty dose of wormwood, first smear the rim of the cup with a sweet coat of yellow honey. The children, too young as yet for foresight, are lured by the sweetness at their lips into swallowing the bitter draught. So they are tricked but not trapped, for the treatment restores them to health. In the same way our doctrine often seems unpalatable to those who have not sampled it, and the multitude shrink from it. That is why I have tried to administer it to you in the dulcet strains of poesy, coated with the sweet honey of the Muses. My object has been to engage your mind with my verses while you gain insight into the nature of the universe and the pattern of its architecture. [...]

Learn, therefore, that *the universe is not bounded in any direction.* [...]

... [N]o rest is given to the atoms, because there is no bottom where they can accumulate and take up their abode. Things go on happening all the time through ceaseless movement in every direction; and atoms of matter bouncing up from below are supplied out of the infinite. [...]

Certainly the atoms did not post themselves purposefully in due order by an act of intelligence, nor did they stipulate what movements each should perform. As they have been rushing everlastingly throughout all space in their myriads, undergoing a myriad changes under the disturbing impact of collisions, they have experienced every variety of movement and conjunction till they have fallen into the particular pattern by which this world of ours is constituted. [...]

It clearly follows that no rest is given to the atoms in their course through the depths of space. Driven along in an incessant but variable movement, some of them bounce far apart after a collision while others recoil only a short distance from the impact. From those that do not recoil far, being driven into a closer union and held there by the entanglement of their own interlocking shapes, are composed firmly rooted rock, the stubborn strength of steel and the like. Those others that move freely through larger tracts of space, springing far apart and carried far by the rebound – these provide for us thin air and blazing sunlight. Besides these, there are many other atoms at large in empty space which have been thrown out of compound bodies and have nowhere even been granted admittance so as to bring their motions into harmony.

This process, as I might point out, is illustrated by an image of it that is continually taking place before our very eyes. Observe what happens when sunbeams are admitted into a building and shed light on its shadowy places. You will see a multitude of tiny particles mingling in a multitude of ways in the empty space within the light of the beam, as though contending in everlasting conflict, rushing into battle rank upon rank with never a moment's pause in a rapid sequence of unions and disunions. From this you may picture what it is for the atoms to be perpetually tossed about in the illimitable void. To some extent a small thing may afford an illustration and an imperfect image of great things. Besides, there is a further reason why you should give your mind to these particles that are seen dancing in a sunbeam: their dancing is an actual indication of underlying movements of matter that are hidden from our sight. There you will see many particles under the impact of invisible blows changing their course and driven back upon their tracks, this way and that, in all directions. You must understand that they all derive this restlessness from the atoms. It originates with the atoms, which move of themselves. Then those small compound bodies that are least removed from the impetus of the atoms are set in motion by the impact of their invisible blows and in turn cannon against slightly larger bodies. So the movement mounts up from the atoms and gradually emerges to the level of our senses, so that those bodies are in motion that we see in sunbeams, moved by blows that remain invisible.[...]

In this connexion there is another fact that I want you to grasp. *When the atoms are travelling straight down through empty space by their own weight, at quite indeterminate times and places they swerve ever so little from their course,* just so much that you can call it a change of direction. If it were not for this swerve, everything would fall downwards like rain-drops through the abyss of space. No collision would take place and no impact of atom on atom would be created. Thus nature would never have created anything. [...]

And now let us turn to a new theme – *the characteristics of the atoms of all substances, the extent to which they differ in shape and the rich multiplicity of their forms.* Not that there are not many of the same shape, but they are by no means all identical with one another. [...]

...You cannot suppose that atoms of the same shape are entering our nostrils when stinking corpses are roasting as when the stage is freshly sprinkled with saffron of Cilicia and a near-by altar exhales the perfumes of the Orient. You cannot attribute the same composition to sights that feast the eye with colour and to those that make it smart and weep or that appear loathsome and repulsive through sheer ugliness. Nothing that gratifies the senses is ever without a certain smoothness of the constituent atoms. Whatever, on the other hand, is painful and harsh is characterized by a certain roughness of matter. Besides these there are some things that are not properly regarded as smooth but yet are not jagged with barbed spikes. These are characterized instead by slightly jutting ridges such as tickle the senses rather than hurt them.... For touch and nothing but touch (by all that men call holy!) is the essence of all our bodily sensations....

Again, things that seem to us hard and stiff must be composed of deeply indented and hooked atoms and held firm by their intertangling branches. In the front rank of this class stand diamonds, with their steadfast indifference to blows. Next come stout flints and stubborn steel and bronze that stands firm with shrieking protest when the bolt is shot. Liquids, on the other hand, must owe their fluid consistency to component atoms that are smooth and round. [...]

You see that many objects are possessed of colour and taste together with smell. Their component matter must therefore be multiform. For scent penetrates the human frame where tint does not go; tint creeps into the senses by a different route from taste. So you may infer that they differ in their atomic forms. Different shapes therefore combine in a single mass, and objects are composed of a mixture of seeds. Consider how in my verses, for instance, you see many letters common to many words; yet you must admit that different verses and words are composed of different letters. Not that there is any lack of letters common to several words, or that there are no two words composed of precisely the same letters; but they do not all alike consist of exactly the same components. So in other things, although many atoms are common to many substances, yet these substances may still differ in their composition....

[...] Granted, then, that empty space extends without limit in every direction and that seeds innumerable in number are rushing on countless courses through an unfathomable universe under the impulse of perpetual motion, *it is in the highest degree unlikely that this earth and sky is the only one to have been created* and that all those particles of matter outside are accomplishing nothing. This follows from the fact that our world has been made by nature through the spontaneous and casual collision and the multifarious, accidental, random and purposeless congregation and coalescence of atoms whose suddenly formed combinations could serve on each occasion as the starting-point of substantial fabrics – earth and sea and sky and the races of living creatures. On every ground, therefore, you must admit that there exist elsewhere other congeries of matter similar to this one which the ether clasps in ardent embrace.

When there is plenty of matter in readiness, when space is available and no cause or circumstance impedes, then surely things must be wrought and effected. You have a store of atoms that could not be reckoned in full by the whole succession of living creatures. You have the same natural force to congregate them in any place precisely as they have been congregated here. You are bound therefore to acknowledge that in other regions there are other earths and various tribes of men and breeds of beasts. [...]

Even if I knew nothing of the atoms, I would venture to assert on the evidence of the celestial phenomena themselves, supported by many other arguments, that the universe was certainly not created for us by divine power: it is so full of imperfections. [...]

...I return now to the childhood of the world, to consider what fruits the tender fields of earth in youthful parturition first ventured to fling up into the light of day and entrust to the fickle breezes.

First of all, the earth girdled its hills with a green glow of herbage, and over every plain the meadows gleamed with verdure and with bloom. Then trees of every sort were given free rein to join in an eager race for growth into the gusty air. As feathers, fur and bristles are generated at the outset from the bodies of winged and four-footed creatures, so then *the new-born earth first flung up herbs and shrubs. Next in order it engendered the various breeds of mortal creatures*, manifold in mode of origin as in form. The animals cannot have fallen from the sky, and those that live on land cannot have emerged from the briny gulfs. We are left with the conclusion that the name of mother has rightly been bestowed on the earth, since out of the earth everything is born. [...]

In those days the earth attempted also to produce a host of monsters, grotesque in build and aspect – hermaphrodites, halfway between the sexes yet cut off from either, creatures bereft of feet or dispossessed of hands, dumb, mouthless brutes, or eyeless and blind, or disabled by the adhesion of their limbs to the trunk, so that they could neither do anything nor go anywhere nor keep out of harm's way nor take what they needed. These and other such *monstrous and misshapen births were created. But all in vain.* Nature debarred them from increase....

In those days, again, *many species must have died out altogether* and failed to reproduce their kind. Every species that you now see drawing the breath of life has been protected and preserved from the beginning of the world either by cunning or by prowess or by speed. In addition, there are many that survive under human protection because their usefulness has commended them to our care. [...]

1.4 Galen, *On Anatomical Procedure*, from *Greek Medicine*, trans. and ed. Arthur J. Brock (London and Toronto: Dent, 1929), pp. 160–5

What tent-poles are to tents, and walls to houses, so to animals is their bony structure; the other parts adapt themselves to this, and change with it. Thus, if an animal's cranium is round, its brain must be the same; or, again, if it is oblong, then the animal's brain must also be oblong. If the jaws are small, and the face as a whole roundish, the muscles of these parts will also necessarily be small; and similarly, if the jaws are prominent, the animal's face as a whole will be long, as also the facial muscles. Consequently also the monkey...is of all animals the likest to man in its viscera, muscles, arteries, veins, and nerves..., because it is so also in the form of its bones. From the nature of these it walks on two legs, uses its front limbs as hands, has the flattest breast-bone of all quadrupeds, collar-bones like those of a man, a round face and a short neck. And these being similar, the muscles cannot be different; for they are extended on the outside of the bones in such a manner that they resemble them in size and form. To the muscles, again, correspond the arteries, veins, and nerves; so these, being similar, must correspond to the bones.

First of all, then, I would ask you to make yourself well acquainted with the human bones, and not to look on this as a matter of secondary importance. Nor must you merely read the subject up in one of these books which are called by some 'Osteology', by others 'The Skeleton', and by others simply 'On Bones', as is my own book; which, by the way, I am certain is better than any previously written, both as regards the exactitude of its matter and the brevity and clearness of its explanations. Make it your earnest business, then, not only to learn exactly from the book the appearance of each of the bones, but to become yourself by the use of your own eyes an eager first-hand observer of human osteology.

At Alexandria this is very easy, since the physicians in that country accompany the instruction they give to their students with opportunities for personal inspection (*autopsia*). Hence you must try to get to Alexandria for this reason alone, if for no other. But if you cannot manage this, still it is not impossible to obtain a view of human bones. Personally I have very often had a chance to do this where tombs or monuments have become broken up. On one occasion a river, having risen to the level of a grave which had been carelessly constructed a few months previously, easily disintegrated this; then by the force of its current it swept right over the dead man's body, of which the flesh had already putrefied, while the bones were still closely attached to one another. This it carried away downstream for the distance of a league, till, coming to a lake-like stretch with sloping banks, it here deposited the corpse. And here the latter lay ready for inspection, just as though prepared by a doctor for his pupil's lesson.

Once also I examined the skeleton of a robber, lying on a mountainside a short distance from the road. This man had been killed by some traveller whom he had attacked, but who had been too quick for him. None of the inhabitants of the district would bury him; but in their detestation of him they were delighted when his body was eaten by birds of prey; the latter, in fact, devoured the flesh in two days and left the skeleton ready, as it were, for anyone who cared to enjoy an anatomical demonstration.

As regards yourself, then, even if you do not have the luck to see anything like this, still you can dissect an ape, and learn each of the bones from it, by carefully removing the flesh. For this purpose you must choose the apes which most resemble man. Such are those in whom the jaws are not prominent nor the canine teeth large. In such apes you will also find the other parts as in man, whence they walk and run on two legs. Those of them, again, that are like the dog-headed baboons... have longer muzzles and large canine teeth; they have difficulty in standing upright on two legs, let alone walking about or running.

But even those apes most like human beings fall somewhat short of the absolutely erect posture. In them the head of the femur is adjusted somewhat obliquely to the hip-socket, and certain of the muscles which run down to the tibia come far forward; both of these factors impair or prevent assumption of the erect posture, as also do their feet, for in these the heels are somewhat narrow and the toes widely separated from each other. These, however, are small matters, and so the ape comes very near being able to stand erect.

Those monkeys which resemble the dog-faced baboons and have a marked divergence from the human type show also as clear difference in their bones. Choose, therefore, among the monkeys the most men-like (*anthropoid*), and learn accurately on them the nature of the bones, comparing them with my writings. You will also have to accustom yourself without delay to their names, as these will also be useful in learning the anatomy of other parts.

Thus if you should also later meet with a human skeleton, you would easily recognise and remember everything. If, on the other hand, you content yourself with reading only, without familiarising yourself beforehand with the sight of the ape's bones, you will neither recognise accurately a human skeleton if you suddenly see one, nor will you be able to remember it; for in order to keep in mind sensible facts, constant familiarity is needed. This is the reason why also among men we quickly recognise those with whom we have often had to do, whereas if we see a person only once or twice and meet him again after any length of time, we pass him by, failing entirely to recognise him or even to remember our previous meeting. Consequently also the far-famed 'fortuitous anatomy' to which some physicians pay honour is not an adequate way of learning the nature of the parts visible. We must have plenty of leisure to inspect each of the parts first, so that, when suddenly seen, it may be recognised. And this should be done preferably with actual human beings, and, if not with them, then with animals like them. [...]

...We must, as I say, study all the bones, either in the human body or, if possible, in the ape, but preferably in both. Then next we proceed to dissect the muscles. For these two parts [bones and muscles] underlie all the others, like foundation-stones. After these you may learn whatever you like first, whether arteries, veins, or nerves. While engaged in the dissection of these you will also learn about the [solid] viscera. Next will come a knowledge of the intestines, of fatty tissue, and of glands; and these, again, must be each examined separately in greater detail. [...]

When apes are not available, one should be prepared to dissect the bodies of other animals, distinguishing at once in what ways they differ from apes. These differences I shall also point out.

[...] Not long ago also I wrote a special anatomy of the muscles....

Whoever wishes, then, can practise from this book, during his dissection of the ape. He will also learn in further detail from this what procedure is needed in dissecting each special part of the muscles.

Practise first on a dead body, in order to know the origin and insertion of each muscle, and also to see whether all its longitudinal fibres have the same appearance or differ in structure. Thus you will find in some muscles the fibres single, in others double, so that they look like several muscles superimposed, and others again have their fibres the reverse of longitudinal. All these observations will be useful to you in surgery as well as in the investigation of functions. For in surgery we are sometimes forced to cut even the muscles themselves, on account of deep-lying abscesses, deposits, or suppurations. And it is most useful to learn the function, so that in the case of extensive injuries..., where a whole muscle has been cut across, we may know beforehand what function is lost; by forecasting this you will avoid being blamed by censorious individuals (who are in the habit of referring loss of function to the doctor's treatment and not to the original injury).

1.5 Galen, *On the Natural Faculties*, trans. A. J. Brock, Loeb Classical Library (London: Heinemann and Cambridge, MA: Harvard University Press, 1963), pp. 56–7, 59, 61

The fact is that those who are enslaved to their sects are not merely devoid of all sound knowledge, but they will not even stop to learn! Instead of listening, as they ought, to the reason why liquid can enter the bladder through the ureters, but is unable to go back again the same way, – instead of admiring Nature's artistic skill – they refuse to learn; they even go so far as to scoff, and maintain that the kidneys, as well as many other things, have been made by Nature *for no purpose*![1] And some of them who had allowed themselves to be shown the ureters coming from the kidneys and becoming implanted in the bladder, even had the audacity to say that these also existed for no purpose; and others said that they were spermatic ducts, and that this was why they were inserted into the neck of the bladder and not into its cavity. When, therefore, we had demonstrated to them the real spermatic ducts entering the neck of the bladder lower down than the ureters, we supposed that, if we had not done so before, we would now at least draw them away from their false assumptions, and convert them forthwith to the opposite view. But even this they presumed to dispute, and said that it was not to be wondered at that the semen should remain longer in these latter ducts, these being more constricted, and that it should flow quickly down the ducts which came from the kidneys, seeing that these were well dilated. We were, therefore, further compelled to show them, in a still living animal, the urine plainly running out through the ureters into the bladder; even thus we hardly hoped to check their nonsensical talk.

Now the method of demonstration is as follows. One has to divide the peritoneum in front of the ureters, then secure these with ligatures, and next, having bandaged up the animal, let him go (for he will not continue to urinate). After this one loosens the external bandages and shows the bladder empty and the ureters quite full and distended – in fact almost on the point of rupturing; on removing the ligature from them, one then plainly sees the bladder becoming filled with urine.

When this has been made quite clear, then, before the animal urinates, one has to tie a ligature round his penis and then to squeeze the bladder all over; still nothing goes back through the ureters to the kidneys. Here, then, it becomes obvious that not only in a dead animal, but in one which is still living, the ureters are prevented from receiving back the urine from the bladder. These observations having been made, one now loosens the ligature from the animal's penis and allows him to urinate, then again ligatures one of the ureters and leaves the other to discharge into the bladder. Allowing, then, some time to elapse, one now demonstrates that the ureter which was ligatured is obviously full and distended on the side next to the

1 Direct denial of Aristotle's dictum that 'Nature does nothing in vain'.

kidneys, while the other one – that from which the ligature had been taken – is itself flaccid, but has filled the bladder with urine. Then, again, one must divide the full ureter, and demonstrate how the urine spurts out of it, like blood in the operation of venesection; and after this one cuts through the other also, and both being thus divided, one bandages up the animal externally. Then when enough time seems to have elapsed, one takes off the bandages; the bladder will now be found empty, and the whole region between the intestines and the peritoneum full of urine, as if the animal were suffering from dropsy. Now, if anyone will but test this for himself on an animal, I think he will strongly condemn the rashness of Asclepiades, and if he also learns the reason why nothing regurgitates from the bladder into the ureters, I think he will be persuaded by this also of the forethought and art shown by Nature in relation to animals. [...]

1.6 Sacrobosco, *The Sphere*, trans. Lynn Thorndike in *The Sphere of Sacrobosco and Its Commentators* (Chicago, IL: University of Chicago Press, 1949), pp. 118–22

The sphere... is divided into the ninth sphere, which is called the 'first moved' or the *primum mobile*; and the sphere of the fixed stars, which is named the 'firmament'; and the seven spheres of the seven planets, of which some are larger, some smaller, according as they the more approach, or recede from, the firmament. Wherefore, among them the sphere of Saturn is the largest, the sphere of the moon the smallest....

[...] The Four Elements. – The machine of the universe is divided into two, the ethereal and the elementary region. The elementary region, existing subject to continual alteration, is divided into four. For there is earth, placed, as it were, as the center in the middle of all, about which is water, about water air, about air fire, which is pure and not turbid there and reaches to the sphere of the moon, as Aristotle says in his book of *Meteorology*. For so God, the glorious and sublime, disposed. And these are called the 'four elements' which are in turn by themselves altered, corrupted and regenerated. The elements are also simple bodies which cannot be subdivided into parts of diverse forms and from whose commixture are produced various species of generated things. Three of them, in turn, surround the earth on all sides spherically, except in so far as the dry land stays the sea's tide to protect the life of animate beings. All, too, are mobile except earth, which, as the center of the world, by its weight in every direction equally avoiding the great motion of the extremes, as a round body occupies the middle of the sphere.

The Heavens. – Around the elementary region revolves with continuous circular motion the ethereal, which is lucid and immune from all variation in its immutable essence. And it is called 'Fifth Essence' by the philosophers. Of which there are nine spheres, as we have just said: namely, of the moon, Mercury, Venus, the sun, Mars, Jupiter, Saturn, the fixed stars, and the last heaven. Each of these spheres incloses its inferior spherically.

THEIR MOVEMENTS. – ... [T]he first movement [of the outermost sphere] carries all the others with it in its rush about the earth once within a day and night, although they strive against it, as in the case of the eighth sphere one degree in a hundred years. This second movement is divided through the middle by the zodiac, under which each of the seven planets has its own sphere, in which it is borne by its own motion, contrary to the movement of the sky, and completes it in varying spaces of time – in the case of Saturn in thirty years, Jupiter in twelve years, Mars in two, the sun in three hundred and sixty-five days and six hours, Venus and Mercury about the same, the moon in twenty-seven days and eight hours.

REVOLUTION OF THE HEAVENS FROM EAST TO WEST. – That the sky revolves from east to west is signified by the fact that the stars, which rise in the east, mount gradually and successively until they reach mid-sky and are always at the same distance apart, and, thus maintaining their relative positions, they move toward their setting continuously and uniformly. Another indication is that the stars near the North Pole, which never set for us, move continuously and uniformly, describing their circles about the pole, and are always equally near or far from one another. Wherefore, from those two continuous movements of the stars, both those that set and those which do not, it is clear that the firmament is moved from east to west.

THE HEAVENS SPHERICAL. – There are three reasons why the sky is round: likeness, convenience, and necessity. Likeness, because the sensible world is made in the likeness of the archetype, in which there is neither end nor beginning; wherefore, in likeness to it the sensible world has a round shape, in which beginning or end cannot be distinguished. Convenience, because of all ... bodies the sphere is the largest and of all shapes the round is most capacious. Since largest and round, therefore the most capacious. Wherefore, since the world is all-containing, this shape was useful and convenient for it. Necessity, because if the world were of other form than round – say, trilateral, quadrilateral, or many-sided – it would follow that some space would be vacant and some body without a place, both of which are false ....

[...] THE EARTH A SPHERE. – That the earth, too, is round is shown thus. The signs and stars do not rise and set the same for all men everywhere but rise and set sooner for those in the east than for those in the west; and of this there is no other cause than the bulge of the earth. Moreover, celestial phenomena evidence that they rise sooner for orientals than for westerners. For one and the same eclipse of the moon which appears to us in the first hour of the night appears to orientals about the third hour of the night, which proves that they had night and sunset before we did, of which setting the bulge of the earth is the cause.

FURTHER PROOFS OF THIS. – That the earth also has a bulge from north to south and vice versa is shown thus: To those living toward the north, certain stars are always visible, namely, those near the North Pole, while others which are near the South Pole are always concealed from them. If, then, anyone should proceed from the north southward, he might go so far that the stars which formerly were always visible to him now would tend toward their setting. And the farther south he went, the more they would be moved toward their setting. Again, that same man now could see stars which formerly had always been hidden from him. And the reverse would happen to anyone going from the south northward. The cause of this is simply the bulge of the earth. Again, if the earth were flat from east to west, the stars would rise as soon for westerners as for orientals, which is false. Also, if the earth were flat from north to south and vice versa, the stars which were always visible to anyone would continue to be so wherever he went, which is false. But it seems flat to human sight because it is so extensive.

SURFACE OF THE SEA SPHERICAL. – That the water has a bulge and is approximately round is shown thus: Let a signal be set up on the seacoast and a ship leave port and sail away so far that the eye of a person standing at the foot of the mast can no longer discern the signal. Yet if the ship is stopped, the eye of the same person, if he has climbed to the top of the mast, will see the signal clearly. Yet the eye of a person at the bottom of the mast ought to see the signal better than he who is at the top, as is shown by drawing straight lines from both to the signal. And there is no other explanation of this thing than the bulge of the water....

THE EARTH CENTRAL. – That the earth is in the middle of the firmament is shown thus. To persons on the earth's surface the stars appear of the same size whether they are in mid-sky or just rising or about to set, and this is because the earth is equally distant from them. For if the earth were nearer to the firmament in one direction than in another, a person at that point of the earth's surface which was nearer to the firmament would not see half of the heavens. But this is contrary to Ptolemy and all the philosophers, who say that, wherever man lives, six signs rise and six signs set, and half of the heavens is always visible and half hid from him. [...]

THE EARTH IMMOBILE. – That the earth is held immobile in the midst of all, although it is the heaviest, seems explicable thus. Every heavy thing tends toward the center. Now the center is a point in the middle of the firmament. Therefore, the earth, since it is heaviest, naturally tends toward that point. Also, whatever is moved from the middle toward the circumference ascends. Therefore, if the earth were moved from the middle toward the circumference, it would be ascending, which is impossible. [...]

Chapter Two
Copernicus and his Revolution

2.1 Copernicus, *On the Revolutions of the Heavenly Spheres*, 1543: (a) Dedication, (b) Foreword by Andreas Osiander, (c) Letter of Nicholas Schönberg, 1536, (d) Preface to the Pope, (e) Introduction to Book One, from Nicholas Copernicus, *Complete Works*, ed. Edward Rosen, vol. 2 (Baltimore, MD: John Hopkins University Press, 1992), (a) p. xix, (b) p. xx, (c) p. xxi, (d) pp. 3–6, (e) pp. 7–8

(a) Dedication

NICHOLAS COPERNICUS
OF TORUŃ

SIX BOOKS ON
THE REVOLUTIONS OF THE HEAVENLY
SPHERES

Diligent reader, in this work, which has just been created and published, you have the motions of the fixed stars and planets, as these motions have been reconstituted on the basis of ancient as well as recent observations, and have moreover been embellished by new and marvelous hypotheses. You also have most convenient tables, from which you will be able to compute those motions with the utmost ease for any time whatever. Therefore buy, read, and enjoy [this work].

Let no one untrained in geometry enter here.

NUREMBERG
JOHANNES PETREIUS
1543

(b) Foreword by Andreas Osiander

To the Reader
Concerning the Hypotheses of this Work[1]

There have already been widespread reports about the novel hypotheses of this work, which declares that the earth moves whereas the sun is at rest in the center of the universe. Hence certain scholars, I have no doubt, are deeply offended and believe that the liberal arts, which were established long ago on a sound basis, should not be thrown into confusion. But if these men are willing to examine the matter closely, they will find that the author of this work has done nothing blameworthy. For it is the duty of an astronomer to compose the history of the celestial motions through careful and expert study. Then he must conceive and devise the causes of these motions or hypotheses about them. Since he cannot in any way attain to the true causes, he will adopt whatever suppositions enable the motions to be computed correctly from the principles of geometry for the future as well as for the past. The present author has performed both these duties excellently. For these hypotheses need not be true nor even probable. On the contrary, if they provide a calculus consistent with the observations, that alone is enough. Perhaps there is someone who is so ignorant of geometry and optics that he regards the epicycle of Venus as probable, or thinks that it is the reason why Venus sometimes precedes and sometimes follows the sun by forty degrees and even more. Is there anyone who is not aware that from this assumption it necessarily follows that the diameter of the planet at perigee should appear more than four times, and the body of the planet more than sixteen times, as great as at apogee? Yet this variation is refuted by the experience of every age. In this science there are some other no less important absurdities, which need not be set forth at the moment. For this art, it is quite clear, is completely and absolutely ignorant of the causes of the apparent nonuniform motions. And if any causes are devised by the imagination, as indeed very many are, they are not put forward to convince anyone that they are true, but merely to provide a reliable basis for computation. However, since different hypotheses are sometimes offered for one and the same motion (for example, eccentricity and an epicycle for the sun's motion), the astronomer will take as his first choice that hypothesis which is the easiest to grasp. The philosopher will perhaps rather seek the semblance of the truth. But neither of them will understand or state anything certain, unless it has been divinely revealed to him.

Therefore alongside the ancient hypotheses, which are no more probable, let us permit these new hypotheses also to become known, especially since they are admirable as well as simple and bring with them a huge treasure of very skillful observations. So far as hypotheses are concerned, let no one expect anything certain from astronomy, which cannot furnish it, lest

---

1 This foreword by Osiander was originally anonymous.

he accept as the truth ideas conceived for another purpose, and depart from this study a greater fool than when he entered it. Farewell.

(c) Letter of Nicholas Schönberg

Nicholas Schönberg, Cardinal of Capua,
to Nicholas Copernicus, Greetings.

Some years ago word reached me concerning your proficiency, of which everybody constantly spoke. At that time I began to have a very high regard for you, and also to congratulate our contemporaries among whom you enjoyed such great prestige. For I had learned that you had not merely mastered the discoveries of the ancient astronomers uncommonly well but had also formulated a new cosmology. In it you maintain that the earth moves; that the sun occupies the lowest, and thus the central, place in the universe; that the eighth heaven remains perpetually motionless and fixed; and that, together with the elements included in its sphere, the moon, situated between the heavens of Mars and Venus, revolves around the sun in the period of a year. I have also learned that you have written an exposition of this whole system of astronomy, and have computed the planetary motions and set them down in tables, to the greatest admiration of all. Therefore with the utmost earnestness I entreat you, most learned sir, unless I inconvenience you, to communicate this discovery of yours to scholars, and at the earliest possible moment to send me your writings on the sphere of the universe together with the tables and whatever else you have that is relevant to this subject. Moreover, I have instructed Theodoric of Reden to have everything copied in your quarters at my expense and dispatched to me. If you gratify my desire in this matter, you will see that you are dealing with a man who is zealous for your reputation and eager to do justice to so fine a talent. Farewell.

Rome, 1 November 1536

(d) Preface to the Pope

TO HIS HOLINESS, POPE PAUL III,
NICHOLAS COPERNICUS' PREFACE
TO HIS BOOKS ON THE REVOLUTIONS

I can readily imagine, Holy Father, that as soon as some people hear that in this volume, which I have written about the revolutions of the spheres of the universe, I ascribe certain motions to the terrestrial globe, they will shout that I must be immediately repudiated together with this belief. For I am not so enamored of my own opinions that I disregard what others may think of them. I am aware that a philosopher's ideas are not subject to the judgement of ordinary persons, because it is his endeavor to seek the truth in all things, to the extent permitted to human reason by God. Yet I hold that completely erroneous views should be shunned. Those who know that the consensus of many centuries has sanctioned the conception that the earth remains at rest in the middle of the heaven as its center would, I reflected, regard it as an insane pronouncement if I made the opposite assertion that the earth moves. Therefore I debated with myself for a long time whether to publish the volume which I wrote to prove the earth's motion or rather to follow the example of the Pythagoreans and certain others, who used to transmit philosophy's secrets only to kinsmen and friends, not in writing but by word of mouth, as is shown by Lysis' letter to Hipparchus. And they did so, it seems to me, not, as some suppose, because they were in some way jealous about their teachings, which would be spread around; on the contrary, they wanted the very beautiful thoughts attained by great men of deep devotion not to be ridiculed by those who are reluctant to exert themselves vigorously in any literary pursuit unless it is lucrative; or if they are stimulated to the non-acquisitive study of philosophy by the exhortation and example of others, yet because of their dullness of mind they play the same part among philosophers as drones among bees. When I weighed these considerations, the scorn which I had reason to fear on account of the novelty and unconventionality of my opinion almost induced me to abandon completely the work which I had undertaken.

But while I hesitated for a long time and even resisted, my friends drew me back. Foremost among them was the cardinal of Capua, Nicholas Schönberg, renowned in every field of learning. Next to him was a man who loves me dearly, Tiedemann Giese, bishop of Chelmno, a close student of sacred letters as well as of all good literature. For he repeatedly encouraged me and, sometimes adding reproaches, urgently requested me to publish this volume and finally permit it to appear after being buried among my papers and lying concealed not merely until the ninth year but by now the fourth period of nine years. The same conduct was recommended to me by not a few other very eminent scholars. They exhorted me no longer to refuse, on account of the fear which I felt, to make my work available for the general use of students of astronomy. The crazier my doctrine of the earth's motion now appeared to most people, the argument ran, so much the more admiration and thanks would it gain after they saw the publication of my writings dispel the fog of absurdity by most luminous proofs. Influenced therefore by these persuasive men and by this hope, in the end I allowed my friends to bring out an edition of the volume, as they had long besought me to do.

However, Your Holiness will perhaps not be greatly surprised that I have dared to publish my studies after devoting so much effort to working them out that I did not hesitate to put down my thoughts about the earth's motion in written form too. But you are rather waiting to hear from me how it occurred to me to venture to conceive any motion of the earth, against the traditional opinion of astronomers and almost against common sense. I have accordingly no desire to conceal from Your Holiness that I was impelled to consider a different system of deducing the motions of the universe's spheres for no other reason than the realization that astronomers do not agree among themselves in their investigations of this subject. For, in the first place, they are so uncertain about the motion of the sun and moon that they cannot establish and observe a constant length even for the tropical year. Secondly, in determining the motions not only of these bodies but also of the other five planets, they do not use the same principles, assumptions, and explanations of the apparent revolutions and motions. For while some employ only homocentrics, others utilize eccentrics and epicycles, and yet they do not quite reach their goal. For although those who put their faith in homocentrics showed that some nonuniform motions could be compounded in this way, nevertheless by this means they were unable to obtain any incontrovertible result in absolute agreement with the phenomena. On the other hand, those who devised the eccentrics seem thereby in large measure to have solved the problem of the apparent motions with appropriate calculations. But meanwhile they introduced a good many ideas which apparently contradict the first principles of uniform motion. Nor could they elicit or deduce from the eccentrics the principal consideration, that is, the structure of the universe and the true symmetry of its parts. On the contrary, their experience was just like some one taking from various places hands, feet, a head, and other pieces, very well depicted, it may be, but not for the representation of a single person; since these fragments would not belong to one another at all, a monster rather than a man would be put together from them. Hence in the process of demonstration or 'method', as it is called, those who employed eccentrics are found either to have omitted something essential or to have admitted something extraneous and wholly irrelevant. This would not have happened to them, had they followed sound principles. For if the hypotheses assumed by them were not false, everything which follows from their hypotheses would be confirmed beyond any doubt. Even though what I am now saying may be obscure, it will nevertheless become clearer in the proper place.

For a long time, then, I reflected on this confusion in the astronomical traditions concerning the derivation of the motions of the universe's spheres. I began to be annoyed that the movements of the world machine, created for our sake by the best and most systematic Artisan of all, were not understood with greater certainty by the philosophers, who otherwise examined so precisely the most insignificant trifles of this world. For this reason I undertook the task of rereading the works of all the philosophers which I could obtain to learn whether anyone had ever proposed other motions of the universe's spheres than those expounded by the teachers of astronomy in the schools. And in fact first I found in Cicero that Hicetas supposed the earth to move. Later I also discovered in Plutarch that certain others were of this opinion. I have decided to set his words down here, so that they may be available to everybody:

> Some think that the earth remains at rest. But Philolaus the Pythagorean believes that, like the sun and moon, it revolves around the fire in an oblique circle. Heraclides of Pontus and Ecphantus the Pythagorean make the earth move, not in a progressive motion, but like a wheel in a rotation from west to east about its own center.

Therefore, having obtained the opportunity from these sources, I too began to consider the mobility of the earth. And even though the idea seemed absurd, nevertheless I knew that others before me had been granted the freedom to imagine any circles whatever for the purpose of explaining the heavenly phenomena. Hence I thought that I too would be readily permitted to ascertain whether explanations sounder than those of my predecessors could be found for the revolution of the celestial spheres on the assumption of some motion of the earth.

Having thus assumed the motions which I ascribe to the earth later on in the volume, by long and intense study I finally found that if the motions of the other planets are correlated with the orbiting of the earth, and are computed for the revolution of each planet, not only do their phenomena follow therefrom but also the order and size of all the planets and spheres, and heaven itself is so linked together that in no portion of it can anything be shifted without disrupting the remaining parts and the universe as a whole. Accordingly in the arrangement of the volume too I have adopted the following order. In the first book I set forth the entire distribution of the spheres together with the motions which I attribute to the earth, so that this book contains, as it were, the general structure of the universe. Then in the remaining books I correlate the motions of the other planets and of all the spheres with the movement of the earth so that I may thereby determine to what extent the motions and appearances of the other planets and spheres can be saved if they are correlated with the earth's motions. I have no doubt that acute and learned astronomers will agree with me if, as this discipline especially requires, they are willing to examine and consider, not superficially but thoroughly, what I adduce in this volume in proof of these matters. However, in order that the educated and uneducated alike may see that I do not run away from the judgement of anybody at all, I have preferred dedicating my studies to Your Holiness rather than to anyone else. For even in this very remote corner of the earth where I live you are considered the highest authority by virtue of the loftiness of your office and your love for all literature and astronomy too. Hence by your prestige and judgement you can easily suppress calumnious attacks although, as the proverb has it, there is no remedy for a backbite.

Perhaps there will be babblers who claim to be judges of astronomy although completely ignorant of the subject and, badly distorting some passage of Scripture to their purpose, will dare to find fault with my undertaking and censure it. I disregard them even to the extent of despising their criticism as unfounded. For it is not unknown that Lactantius, otherwise an illustrious writer but hardly an astronomer, speaks quite childishly about the earth's shape, when he mocks those who declared that the earth has the form of a globe. Hence scholars need not be surprised if any such persons will likewise ridicule me. Astronomy is written for astronomers. To them my work too will seem, unless I am mistaken, to make some contribution also to the Church, at the head of which Your Holiness now stands. For not so long ago under Leo X the Lateran Council considered the problem of reforming the ecclesiastical calendar. The issue remained undecided then only because the lengths of the year and month and the motions of the sun and moon were regarded as not yet adequately measured. From that time on, at the suggestion of that most distinguished man, Paul, bishop of Fossombrone, who was then in charge of this matter, I have directed my attention to a more precise study of these topics. But what I have accomplished in this regard, I leave to the judgement of Your Holiness in particular and of all other learned astronomers. And lest I appear to Your Holiness to promise more about the usefulness of this volume than I can fulfill, I now turn to the work itself.

(e) Introduction to Book One

Book One

INTRODUCTION

Among the many various literary and artistic pursuits which invigorate men's minds, the strongest affection and utmost zeal should, I think, promote the studies concerned with the most beautiful objects, most deserving to be known. This is the nature of the discipline which deals with the universe's divine revolutions, the asters' motions, sizes, distances, risings and settings, as well as the causes of the other phenomena in the sky, and which, in short, explains its whole appearance. What indeed is more beautiful than heaven, which of course contains all things of beauty? This is proclaimed by its very names [in Latin], *caelum* and *mundus*, the latter denoting purity and ornament, the former a carving. On account of heaven's transcendent perfection most philosophers have called it a visible god. If then the value of the arts is judged by the subject matter which they treat, that art will be by far the foremost which is labeled astronomy by some, astrology by others, but by many of the ancients, the consummation of mathematics. Unquestionably the summit of the liberal arts and most worthy of a free man, it is supported by almost all the branches of mathematics. Arithmetic, geometry, optics, surveying, mechanics and whatever others there are all contribute to it.

Although all the good arts serve to draw man's mind away from vices and lead it toward better things, this function can be more fully performed by this art, which also provides extraordinary intellectual pleasure. For when a man is occupied with things which he sees established in the finest order and directed by divine management, will not the unremitting contemplation of them and a certain familiarity with them stimulate him to the best and to admiration for the Maker of everything, in whom are all happiness and every good? For would not the godly Psalmist [92:4] in vain declare that he was made glad through the work of the Lord and rejoiced in the works of His hands, were we not drawn to the contemplation of the highest good by this means, as though by a chariot?

The great benefit and adornment which this art confers on the commonwealth (not to mention the countless advantages to individuals) are most excellently observed by Plato. In the *Laws*, Book VII, he thinks that it should be cultivated chiefly because by dividing time into groups of days as months and years, it would keep the state alert and attentive to the festivals and sacrifices. Whoever denies its necessity for the teacher of any branch of higher learning is thinking foolishly, according to Plato. In his opinion it is highly unlikely that anyone lacking the requisite knowledge of the sun, moon, and other heavenly bodies can become and be called godlike.

However, this divine rather than human science, which investigates the loftiest subjects, is not free from perplexities. The main reason is that its principles and assumptions, called 'hypotheses' by the Greeks, have been a source of disagreement, as we see, among most of those who undertook to deal with this subject, and so they did not rely on the same ideas. An additional reason is that the motion of the planets and the revolution of the stars could not be measured with numerical precision and completely understood except with the passage of time and the aid of many earlier observations, through which this knowledge was transmitted to posterity from hand to hand, so to say. To be sure, Claudius Ptolemy of Alexandria, who far excels the rest by his wonderful skill and industry, brought this entire art almost to perfection with the help of observations extending over a period of more than four hundred years, so that there no longer seemed to be any gap which he had not closed. Nevertheless very many things, as we perceive, do not agree with the conclusions which ought to follow from his system, and besides certain other motions have been discovered which were not yet known to him. Hence Plutarch too, in discussing the sun's tropical year, says that so far the motion of the heavenly bodies has eluded the skill of the astronomers. For, to use the year itself as an example, it is well known, I think, how different the opinions concerning it have always been, so that many have abandoned all hope that an exact determination of it could be found. The situation is the same with regard to other heavenly bodies.

Nevertheless, to avoid giving the impression that this difficulty is an excuse for indolence, by the grace of God, without whom we can accomplish nothing, I shall attempt a broader inquiry into these matters. For, the number of aids we have to assist our enterprise grows with the interval of time extending from the originators of this art to us. Their discoveries may be compared with what I have newly found. I acknowledge, moreover, that I shall treat many topics differently from my predecessors, and yet I shall do so thanks to them, for it was they who first opened the road to the investigation of these very questions.

2.2 G. J. Rheticus, *Holy Scripture and the Motion of the Earth* [1540], trans. R. Hooykas, from *G. J. Rheticus' Treatise on Holy Scripture and the Motion of the Earth* (Amsterdam: North-Holland Publishing Company, 1984), pp. 91–101

Furthermore, there will not be lacking those who will bellow that it is monstrous to attribute movements to the earth, and who will take occasion to draw on and display their wisdom taken from the philosophers of nature. They are ridiculous, as if God's power could be measured by our force or our intellect. Are we to think that anything is impossible for God, Who, by His Word, made the whole natural order out of nothing? Are we to tie God to the disputations of the Peripatetics about the heavy and the light, Him Who is tied to no place, but Who fills the whole world and is everywhere present and powerful, and Who places the Antipodes firmly on the earth? And He Who decreed that heaven should everywhere be above, is He unable to give the earth natural movements in accord with its shape? Not in any way, in my opinion. Nor therefore is it necessary for us to agree with Aristotle, when he teaches that movements from the centre, to the centre, and around the centre are distinct from each other. We may indeed rightly insist, especially as mathematical reasoning compels us, that the whole earth moves in a circle, and that therefore this motion is present in all its parts. Furthermore, that motions from the centre and to the centre appear by accident in the elements, as when air, confined by water or earth, tends upwards, and breaks out towards the place, which nature has assigned to it, and earth that has been thrown up into the air, keeps falling downward, until it stops on the earth, or occupies the centre of the earth, that being its naturally allotted place. Wherefore we would say that to bring something into its place is nothing else than to bring it into its proper 'Form', and that the ancients rightly held – in opposition to Aristotle – that like is attracted by like. For this is the reason and divine ordinance, on account of which the Sun, the Moon, the stars, and the earth are spherical, *and it is not proved that heavy objects tend to reach the centre of the universe, but only towards the centre of their own globe.* [...]

Now, the passages of Scripture, which occur to us as the principal ones against the mobility of the earth, are roughly these: Isaiah 42: 'Thus says the Lord God who creates the heavens and spreads them out, who settles the earth and the things which grow out of it, giving breath to the people which is on it, and His Spirit to those who walk upon it'. Likewise, chap. 44: 'I am the Lord, who makes all things, spreading out the heavens and settling the earth, I alone and none with Me'. And later, chap. 48: 'My hand also has founded the earth, and My right hand has spanned the heavens'.

David, Ps. 92: 'For He has so established the earth that it shall not be moved'.

Ps. 101: 'In the beginning Thou, Lord, hast founded the earth, and the heavens are the work of Thy hands'.

Ps. 103: 'Who hast founded the earth on its foundations. It will not be shaken for ever'.

Ps. 118: 'Thy verity is from generation to generation; Thou hast founded the earth, and it abides'.

Zacharias, ch. 12: '(Thus) said the Lord who spreads out the heavens, who founds the earth and forms the spirit of man within him'.

On this account, the ancients believed that the earth had been made immobile, and, following Aristotle, they asserted that it was placed at the centre of the universe, where it was to be in its place of rest. But we say that it is not to be taken, as if He created an immobile world. And in support of this our belief we have, in the first place, Mathematics, in the second place, other passages of Scripture. What in heaven is more unfixed or mobile than the Moon? If, however, 'to found' signified 'to make immobile', David would be saying that it [the moon], along with the rest of the stars, was immobile, when he says: 'When I see Thy heavens, the works of Thy fingers, the moon and the stars which Thou hast founded'. In a similar way, God did not render the earth immobile, either by fixing or by establishing it, for Scripture attributes the same to heaven, as when David says: 'By the Word of the Lord were the heavens established, and by the Spirit of His mouth was all their strength (ordained)'. Likewise Solomon: 'The Lord by wisdom has founded the earth, by understanding He has established the heavens'.

Therefore, to us the passage quoted from Ps. 103 unties the entire knot of the discussion. Just as David said that the earth was founded, – that is, fixed and established – , on its foundations, which it is to keep for ever, so we also will correctly understand the Moon, and any other moving heavenly body, to be founded and fixed, as it were, on its stability, from which it will never decline. For it is clear that each of these bodies, by divine ordinance, is maintained in its 'way of being' (as we usually say).

For, although on earth there occur corruptions, generations, and all kinds of alterations, yet the earth itself remains in its wholeness as it was created. Fire, air, water, earth – everything keeps its place and fulfils the task for which it was created. Thus, as whatever Scripture means by the name 'earth', is founded on its own stability, so also are its parts, as the Psalmist bears witness when he says: 'Thou hast established the sea in Thy strengh', etc. But, whether there are changes in the Moon and the other heavenly bodies, or not, I do not see, how we could determine. For, if somebody would live on the Moon, I do not think that he would be able to judge anything about changes on earth. And Nicholas of Cusa, in his 'Learned Ignorance', argues at length that the earth also is luminous and so one of the stars. Furthermore, since motion also belongs to the way of being of the earth and of the other moving bodies, *it should be said that each of them has been founded on its stability, that is, so created, that it maintains its established course, (to use a term of Pliny's), and attains its prescribed positions.* And unless, for the sake of such a stability of motion, these things had been fixed in a definite and perpetual Law, we also would have no certain calculation of time, which God nevertheless wished us to have, as we read in the first chapter of Genesis. From all this it is plain that it cannot be proved from the sacred writings that the earth is immobile. Therefore, he who assumes its mobility in order to bring about a reliable calculation of times and motions, is not acting against Holy Scripture.

But let us come to the testimonies of Scripture concerning the Sun's mobility.

That the sun by its motion, assigned to it by God, is the originator of day and night and of all the changes of the seasons, and also is itself carried in an oblique circle, according to the hypotheses of Ptolemy and the ancients, seems to be proved by the following testimonies. *Genesis* 1: 'And God made two great lights, a greater light to rule the day and a lesser light to rule the night and the stars, and God placed them in the firmament of heaven, that they might shine upon the earth and rule the day and the night, and to divide the light from the darkness'. *Genesis* 19: 'The sun went out over the earth, and Lot went in to Zoar'.

Joshua, ch. 10: 12 *ff*: 'Then Joshua spoke to the Lord in the day in which He delivered up the Amorite in the sight of the children of Israel, and he said in their presence: Sun, do not move over Gibeon and Moon do not move over the valley of Ajalon. And the Sun and the Moon stood still until the people had avenged themselves upon their enemies', and then he adds: 'So the Sun stood still in the midst of heaven, and did not hasten to set for the space of one day. Never before and never since has there been so long a day, as when the Lord obeyed the voice of a man and fought for Israel'.

IV *Kings, ch.* 20, and *Isaiah, ch.* 38, addressing Hezekiah when he was ill: 'Behold, I shall cause the shadow of the lines through which it has gone down on the sundial of Ahaz to return ten lines. And the Sun went back ten lines, by which degrees it has gone down'.

Ecclesiasticus repeats this same passage in ch. 48, speaking about Hezekiah: 'In his days the Sun went back, and added life to the king'.

David, *Psalm* 103: 'He made the Moon for seasons. The Sun knows his setting'. The passage quoted from Genesis is repeated in *Psalm* 135 and in *Jeremiah, ch.* 37.

Baruch, ch. 6: 'The Sun also and the Moon and the stars obey, since they are magnificent and sent forth for their purposes'.

Ecclesiastes, ch. 1: 'The Sun rises and sets and returns to his place, whence rising again he revolves through the south and turns to the north. Illuminating the universe in his course the Spirit goes on and returns along his circles'.

So David in *Psalm* 18: 'He put His tabernacle in the Sun, who as a bridegroom coming from his wedding chamber rejoices like a giant to run the race; his coming forth is from the highest heaven. And his course is up till the highest [of it]; there is nothing which can hide from his heat'. To these testimonies of Scripture concerning the motion of the Sun, the answer is not difficult. We admit that the Sun is the natural source of light, and God's administrator, as the Psalmist says: 'He put His tabernacle in the Sun', in order to lighten the whole of created nature, and sends forth light and it goes, and He calls it back and it obeys Him in awe, as is written in *Baruch, ch.* 3.

Moreover, we do not deny obvious experience, which is that thanks to the Sun we have day, spring, summer and the other seasons of the year. But when we say that we receive these things from the Sun, just as the Moon receives its light, according to its changing relation to the Sun, we do so that it may be known to the learned through the authority of Urania, to whom the lover of truth must defer. – .

Furthermore, neither do we deny the clear words of Scripture, – since it does not assign a daily and a yearly motion to the Sun and, if you would have it so, also does not assign a motion of precession, since from it the seasons, days and years are measured as from a [fixed] point. Surely, we must consider what sort of movements these are. Everything that appears to move does so either because of the motion of the thing itself, or because of the movement of one's vision, or because of the movement of the object, as well as of the centre of vision. Common speech, however, mostly follows the judgment of the senses. Therefore these differences of movement are not distinguished in this from each other. This also, when a point of view determines something, and we know that in fact the matter stands otherwise, as it may often be noticed in everyday speech and in writings as, when following the judgment of our senses, while we sail from the harbour, we say that the land and the towns recede from us, and when navigating we say that the mountains and lands rise up out of the sea, and that the sun and the stars sink into it, and in our speech we do not distinguish the truth from the appearances.

When, however, we think as [persons] who seek the truth about things, we distinguish in our minds between appearance and reality. As the saying goes: we will judge as the few, but speak as the many. Thus when right reason concludes that the Sun is immobile, even though our eyes lead us to think it moves, we do not abandon the accepted way of speaking. We say that the Sun rises and sets, establishes the day and the year by its motion, even though we hold this to be true only in appearance, as our reason concludes to its immobility. In fact, it is the same going north, when we say that the pole [star] rises, because so it seems to us. But reason knows well that it stays fixed, and only seems to grow higher as we see it, because of our own moving towards it.

But it is too well-known to need further proof, that Holy Scripture uses common and received forms and figures of speech. Whence it is clear that, however much we insist on the many descriptions of the sun's movement adduced from Scripture, these are to be understood as referring to its apparent motion, without in any way going beyond the bounds set by St. Augustine, nor introducing anything from which something inconvenient might follow. Therefore the texts of Scripture concerning the Sun's movement, which seem to argue against us, will not turn out to be at variance with the best verified results of the recent restoration of astronomy. [. . .]

2.3 Andreas Vesalius, Preface, *On the Fabric of the Human Body*, 1543, trans. W. F. Richardson and J. B. Carman (San Francisco, CA: Norman Publishing, 1998), pp. l–lvii

. . . Anatomy is an important part of natural philosophy; to it, since it embraces the study of man and must properly be regarded as the prime foundation of the whole art of medicine and the source of everything that constitutes it, Hippocrates and Plato attributed such importance that they did not hesitate to ascribe to it first place among the component parts of medicine. Previously this study was uniquely pursued by physicians, who strained every nerve in the process of mastering it; but when they handed over the task of surgery to others they lost the art of dissection, and this meant that the whole of anatomy went forthwith into a sad decline. For so long as the physicians declared that the treatment only of internal afflictions was their province, they considered that knowledge of the viscera was all that they required, and they neglected the fabric of the bones and muscles, and of the nerves, veins, and arteries that permeate the bones and muscles, as if it were none of their business. Furthermore, when the whole practice of cutting was handed over to the barbers, not only did the physicians lose firsthand knowledge of the viscera but also the whole art of dissecting fell forthwith into oblivion, simply because the physicians would not undertake to perform it, while they to whom the art of surgery was entrusted were too unlettered to understand the writings of the professors of anatomy.[1] It is quite impossible that such people should preserve for us that most difficult and abstruse art which had been handed over to them; nor is it possible to prevent that evil fragmentation of the healing art from importing into our Colleges that detestable ritual whereby one group performs the actual dissection of a human body and another gives an account of the parts: the latter aloft on their chairs croak away with consummate arrogance like jackdaws about things that they have never done themselves but which they commit to memory from the books of others or which they expound to us from written descriptions, and the former are so unskilled in languages that they cannot explain to the spectators what they have dissected but hack things up for display following the instructions of a physician who has never set his hand to the dissection of a body but has the cheek to play the sailor from a textbook. So the teaching in our colleges is all wrong, and days are frittered away in ridiculous inquiries; a butcher in

1 Vesalius frequently uses this phrase to refer to the ancient writers on anatomy.

a shambles could teach a practitioner more than the spectators are shown amidst all this racket. There are even some colleges (I shall not name them) where virtually no thought is ever given to dissecting the human structure. So much did the ancient art of medicine decline many years ago from its former glory.

In the present age, however, which by the will of the gods is subject to your Majesty's wise rule, things have taken a turn for the better, and medicine, along with all other studies, has begun so to come to life again and to raise its head from the profound darkness which enveloped it that in several universities it has beyond all argument come close to recovering its former glory. Nothing was more urgently required than knowledge of the parts of the human body, a knowledge that had become almost extinct; and therefore I, challenged by the example of so many outstanding men, decided to lend such assistance as I could by whatever means were available to me. Not wishing to be the only one who remained idle while everyone else was achieving such success for the benefit of common studies, or to let down my forebears, I decided to recall this branch of natural philosophy from the dead. My hope was that, though it would be impossible to treat this subject more perfectly in our time than in the time of any of the earlier professors of anatomy, yet at least the point might eventually be reached where one need not hesitate to assert that the modern knowledge of anatomy was comparable with the ancient, and that in this our time nothing had suffered a more serious collapse and made a more notable recovery than the study of anatomy.

This undertaking would never have been brought to a successful conclusion had I not, while studying medicine in Paris, personally set my hand to the task by attending one or two public dissections and acquiescing in the careless and superficial demonstration to myself and my fellow students of a few viscera by unskilled barbers. So perfunctorily was anatomy treated in Paris (where we first saw the successful rebirth of medicine) that I was moved to take matters into my own hands. Having been involved in a number of animal dissections under the famous Jacobus Sylvius (who can never be praised too highly), I was induced by the exhortations of my fellow students and my teachers to take public charge of the third dissection I ever attended and to perform it more thoroughly than was the custom. The next time I took it on (the barbers having now been dismissed from the task) I tried to demonstrate the muscles of the hand and to carry out a more accurate dissection of the viscera; apart from eight abdominal muscles (hacked up badly and in the wrong order) I had never been shown any muscle or any bone, much less an accurate series of nerves, veins or arteries. Then the tumult of war necessitated my return to Louvain, where for eighteen years the physicians had done no anatomy, even in their dreams; and there, with the intention of being of some assistance to the students of that university and of increasing my own expertise in a subject that, though very difficult, is in my opinion essential to the art of medicine, I dissected and expounded the human fabric in greater detail than I had at Paris; and as a result of this the younger professors at that university now diligently devote much serious effort to acquiring knowledge of the parts of the human body because they realize that this knowledge furnishes them with a quite outstanding resource for philosophical inquiry. At Padua, in the world's most famous university, I have continued to devote much effort to my researches into the construction of the human frame; and because the study of anatomy is very relevant to the profession of surgery I have for the last five years been lecturing on surgery. (I undertook this so as not to divorce myself completely from the rest of medicine, and my salary has been paid by the illustrious VENETIAN SENATE, whose generosity to the world of scholarship is without equal.) I have performed frequent dissections both here and in Bologna and, discarding the ridiculous system of the schools, have given both the demonstration and the accompanying commentary myself, while seeking to insure that all the knowledge that has come down to us from the ancient world is made available and that there is no part of the body whose construction is still to seek.

It was the laziness of the medical fraternity that saw to it that the writings of... outstanding anatomists should no longer be available to us. We have not even a fragment of a single page of any of the more than twenty illustrious authors mentioned by Galen in Book II of his commentary on Hippocrates: *The Nature of Man*. Less than half of Galen's *On Anatomical Procedures* has been preserved from death. As to those who came after Galen, in which class I place Oribasius, Theophilus, the Arabs, and all of our own writers whom I have so far read, everything worth reading in their books was borrowed from Galen. I swear that anyone with some experience in dissecting can see that they had never undertaken to dissect a human body. In fact the foremost of them, absorbed in the stylistic quality of their own writing and relying entirely on dissections incompetently carried out by others, produced costly summaries of Galen and did it very badly, refusing to depart a fingernail's breadth from them while they sought to grasp his meaning. They state at the beginning of their books that their writings have been stitched together entirely from the teachings of Galen and that everything they say is Galen's, adding that if anyone finds anything wrong in them Galen must bear part of the blame. Yet such faith did they all have in him that there has never been a single practitioner who believed that even the most trifling error had ever been found in Galen's writings or ever could be found in them. In fact, however, (leaving aside the fact that Galen often corrects himself and, in later books written when he had become more expert, more than once points out some piece of negligence committed in earlier books, and even directly contradicts himself) I am quite certain, on the basis of the art of dissection as now reborn combined with a careful reading of Galen's works and many textual restorations thereof for which I make no apology, that he himself had never cut open a human body and furthermore that, deceived by his apes (although he did chance upon two human skeletons) he frequently and quite wrongly finds fault with the ancient physicians who actually did their training by dissecting human material. For indeed one can find very many instances in Galen where he was wrong even about his apes; not to mention the remarkable fact that, granted the infinite multiplicity of differences between the organs of the human and the simian bodies, he yet noticed none of them except in the digits and the knee joint. Even these he would no doubt have missed if they had not been obvious to him without any need for human dissection.

At this point, however, I have no intention whatever of criticizing the false teachings of Galen, who is easily first among the professors of dissection, for I certainly do not wish to start off by gaining a reputation for impiety toward him, the author of all good things, or by seeming insubordinate to his authority. For I am well aware how upset the practitioners (unlike the followers of Aristotle) invariably become nowadays, when they discover in the course of a single dissection that Galen has departed on two hundred or more occasions from the true description of the harmony, function, and action of the human parts, and how grimly they examine the dissected portions as they strive with all the zeal at their command to defend him. Yet even they, drawn by their love of truth, are gradually calming down and placing more faith in their own not ineffective eyes and reason than in Galen's writings; they are making careful notes of the contradictions, which they have not simply begged from other authors and which are supported by something better than a mere heap of authorities, and are sending the notes to their friends in various places with a firm but friendly exhortation to carry out their own investigation and so gain knowledge of the real anatomy. As a result there is hope that this last will soon be cultivated in all universities as it was once practiced in Alexandria long ago in the days of . . . famous experts in dissection.

I have done my best to bring the assistance of the Muses to this process by setting out afresh our knowledge of the parts of the human body in seven books; this is over and above my other publications on this subject, which certain plagiarists, thinking me well away from Germany, have passed off as their own. The order of these books is that in which I normally treat the subject in the congregation of eminent men in this city and in Bologna. This means that those who were present at my dissections will have notes of what I demonstrated and will be able with greater ease to demonstrate anatomy to others. But the books will be particularly useful also for those who cannot see the real thing, since they consider at sufficient length the number of each part of the body, its position, shape, size, substance, connection with other parts, use, function and many similar matters; all of these are aspects of the nature of the parts into which we normally inquire when dissecting. The method of dissecting the dead and the living is also described, and pictures of all the parts are incorporated into the text of the discourse, so as virtually to set a dissected body before the eyes of students of the works of Nature. [. . .]

I am not unmindful of the opinion of certain people, who strongly deny that even the most exquisite delineations of plants and of parts of the human body should be set before students of the natural world; they take the view that these things should be learned, not from pictures but from careful dissection and examination of the actual objects. In adding to the context of my discourse such detailed diagrams of the parts (and God grant that the printers will not ruin them!) it was never my intention that students should rely on these without ever dissecting cadavers; rather I would, as Galen did, urge students of medicine by every means at my command to undertake dissections with their own hands. If the custom of the ancients, who trained their lads at home in carrying out dissections as much as in writing the alphabet and in reading, had been brought down to the present time, I would be very happy that we, like the ancients, should dispense not only with pictures but with commentaries as well; for the ancients only began to write about anatomical procedures when they decided it was permissible to communicate the art, not only to one's children but also to grown men from other families who were taken on because of their good qualities. As soon as the custom of training lads in dissection was discontinued, forthwith it came about of necessity that they learned anatomy less well, lacking the training that they used to begin in childhood. And so when the art dropped away altogether from the sons of Asclepius and went downhill for many centuries, there was a need of books to preserve its theory untouched.

In fact, illustrations greatly assist the understanding, for they place more clearly before the eyes what the text, no matter how explicitly, describes. This fact is well known in respect of geometry and other branches of mathematics. But in addition our pictures of the parts of the body will give particular pleasure to those people who do not always have the opportunity of dissecting a human body or who, if they do have the opportunity, are by nature so squeamish (a very inappropriate quality in a physician) that, although they are fascinated and delighted by the study of man (which attests, if anything does, to the wisdom of the infinite Creator of the world), yet they cannot bring themselves to the point of ever actually attending a dissection. But in any case I have throughout the work pursued single-mindedly the one aim of giving assistance to as many people as possible in a matter that is extremely recondite and no less arduous, by detailing as accurately and completely as I can the investigation of the fabric of the human body, which is formed, not from ten or twelve different parts (as might appear at a casual glance) but from something like a thousand. I aim also to do something not without value for students of medicine by interpreting those books of Galen which have been preserved to posterity and which, like all the monuments of his divine genius, now need the work of an expositor.

I am not unaware that, because of my age (I am not yet twenty-eight years old), my undertaking will wield little authority and, because I have frequently demonstrated the inaccuracy of Galen's teachings, will not be safe from the attacks of those who, not having attended my anatomical demonstrations or done any proper dissecting of their own, will leap unthinkingly to Galen's defense with a variety of notions. My treatise needs the patronage of some influential public figure if its publication is to take place with good auspices. Because it cannot be more splendidly adorned or more safely protected than by the immortal name of the divine emperor CHARLES, mighty and unconquered, I beg and beseech your imperial Majesty with all reverence as a suppliant that you should allow this my juvenile study, despite its many deficiencies and its many shortcomings, to dwell in the hands of mankind beneath your guidance, splendor, and patronage until such time as, fortified by the experience, the knowledge and the judgment that come with age, I may make it more worthy of a great and powerful prince or may offer a work of significance on some other topic taken from our art. But I suspect that, from . . . the whole of natural philosophy, nothing could be more grateful and acceptable to your Majesty than an account of the research by means of which we make discoveries about the body and the mind, about the divine power that arises from the harmony of both, and hence about our own selves (which is the proper study of mankind). . . .

Chapter Three
The Spread of Copernicanism in Northern Europe

3.1 Robert Recorde, *The Castle of Knowledge*, 1556 (Amsterdam: Theatrum Orbis Terrarum, 1975)

THE PREFACE TO THE
READER

If reasons reache transcende the Skye,
  Why shoulde it then to earthe be bounde?
The witte is wronged and leadde awrye,
  If mynde be maried to the grounde.

THEREFORE,

WHEN SCIPIO BEHELDE OUTE of the high heavens the smallenes of the earth with the kingdomes in it, he coulde no lesse but esteeme the travaile of men moste vaine, which sustaine so muche grief with infinite daungers to get so small a corner of that lyttle balle. So that it yrked him (as he then declared) to considre the smalnes of that their kingdom, whiche men so muche did magnifie. Who soever therefore (by Scipions good admonishment) doth minde to avoide the name of vanitie, and wishe to attayne the name of a man, lette him contemne those trifelinge triumphes, and little esteeme that little lumpe of claye: but rather looke upwarde to the heavens, as nature hath taught him, and not like a beaste go poringe on the grounde, and lyke a seathen swine runne rootinge in the earthe. Yea let him think (as Plato with divers other philosophers dyd trulye affirme) that for this intent were eies geven unto men, that they might with them beholde the heavens: whiche is the theatre of Goddes mightye power, and the chiefe spectakle of al his divine workes. [...]

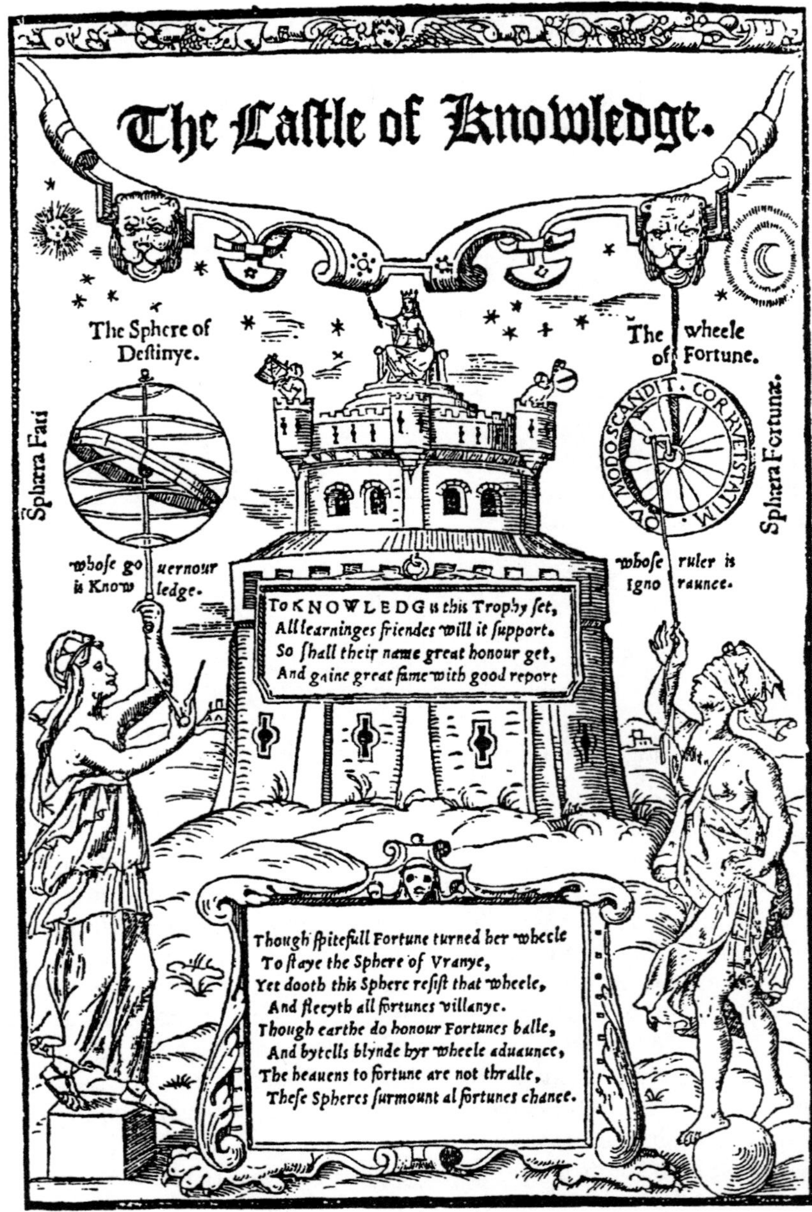
The Castle of Knowledge.
The Sphere of Destinye.
The wheele of Fortune.
Sphæra Fati
Sphæra Fortunæ.
QVI MODO SCANDIT. CORRVET STATIM.
whose gouernour is Knowledge.
whose ruler is Ignoraunce.
To KNOWLEDG is this Trophy set,
All learninges friendes will it support.
So shall their name great honour get,
And gaine great fame with good report
Though spitefull Fortune turned her wheele
To staye the Sphere of Vranye,
Yet dooth this Sphere resist that wheele,
And fleeyth all fortunes villanye.
Though earthe do honour Fortunes balle,
And bytells blynde hyr wheele aduaunce,
The heauens to fortune are not thralle,
These Spheres surmount al fortunes chance.

I mynde not to discourse in declaringe the profite and commodity of Astronomye, but only to admonishe briefly the reader, that hee maye thinke the study woorthye his travaile, and to knowe it to be the moste necessary studye that can be, for anye man that desireth perfection of wisedome. What benefite doth come by it to the true knowledge of husbandrye and navigation, I am assured the verye simplest in those artes do partlye perceave: and the cunningest in the same do so fullye understande, that they judge them selves naked and bare without it, and utterlye destitute of all excellency in their arte. In physicke the use of it is so large in judginge duely of complexions, in prescribinge righte ordre of diete and conversation, in governaunce of healthe, for juste ministration of medicines in time of sickenes, and in righte judgement of the Criticall daies, that without it physicke is to be accompted utterlye imperfecte. For proofe whereof althoughe there be infinite places in Hippocrates and Galene, and divers other good writers, yet hee that hathe readde in Hippocrates but that one booke of Ayer, water, and Regions, and Galen his third boke of Criticall daies, can not be ignoraunte howe necessarye an instrument Astronomy is unto Physicke, as bothe those bookes do testifie at large. But omittinge the testimonies of famous wryters (whiche would make a wonderfull volume of them selves, if they were written only together) I wyll use a simple plaine proofe manifest to all men, and therefore moste apte for to perswade all men. Firste to beginne with sowinge of graine, with grassynge and plantinge, who is so rude, but knoweth that without these be dulye doone, and in their seasonable time, men can not conveniently lyve on the earthe? And howe are their times knowen, but by the risinge and setting of certaine notable starres? Peradventure some man will answere, that by the monethes of the yeare all men do know their times without farther Astronomy. Whiche answere is suche, as if a carpentar or mason shoulde saye, that he can woorke with his compasse, rular, squire, plumbe rule, and suche like instrumentes, without any knowledg in Geometrye. but how ridiculous an answer this were, all men can judge. Likewaies, if a master of a shippe would say, that he can saile and governe his course by his compasse and his carde, with his quadrante and his other instrumentes, without any knowledge in Cosmographye or Astronomye, would not all men that heare him, deryde him, or thinke him madde, for speaking so undiscreatly, especially such as know (as few ar ignorant therein) that all those instrumentes are made by those artes, and appertain to them? So if the distinction of times do depende of Astronomy all together, and the monethes woulde soone runne out of their courses, if the ayde that it hathe by that arte were neglected, so that Michelmas day wold happen in the Spring time, and the Annunciation of our Ladye would fall after harvest (as the truthe is, it would do, if Astronomicall accompte were not) who can shew him selfe so madde as to denye the necessarye use of Astronomye, in due keping the times of the yeares? The ecclesiasticall historye dothe declare at large, and other writers in greate numbre do testifie, that greate controversye hath beene in the churche, for the righte observation of Easter, which controversye could never be decided but by the knowledge of Astronomye. And of late yeares in divers councelles redresse hath beene sought for the juste observation of it: consideringe that if errour be in it, all other moveable feastes, are wrongly kepte by that occasion, and Lente displaced so, that some tyme it hath beene kepte sooner then it ought, and at other times later then it oughte. Whiche faulte can never bee redressed but by astronomy. Whereby it appeareth also manifestly, that in ecclesiasticall maters Astronomy hath a great use, but that is so well knowen, that everye man almoste doth confesse it. And generally who so ever dothe take benefite by the dewe distinction of the yeare, he can not chose but acknowledge that the same commoditie doth come by Astronomy. If I should specially and perticularlye discourse in everye kinde of science and artes, and shewe how they are ayded by astronomye, I should make my preface overlonge, and repeate thinges that all men doth knowe. In lawe for contractes and bargaines the time is most necessarye to be observed: but especiallye if they depende of moveable feastes, wherein astronomy must discusse the doubte. In Grammar, Logike and Rhetorike howe needefull it is, and in histories also, I neede say nothinge, but remitte all men to the readinge of those bokes, which are used in those artes, whereby it shall appeare, that without the principles of Astronomye those bookes can not bee understande. Then for vulgare artes how the knowledge of ebbes and fluddes doth profite, manye men, but speciallye mariners can testifie: and namely suche as understande, what errour commeth by the difference of the true accompte therein and the vulgare accompte. Againe for loppinge of trees and wodde fall, and divers other observations in husbandry, the consideration of the sonne and commonlye of the moone doth greatly healpe. Wherfore I maye conclude, that in all artes and sciences, in lawe, physicke and divinitie, in mariners arte and husbandrye, the profite of Astronomye is exceding necessarye. But above all other things the testimonye of Christe in the scripture doth most approve it, when he doothe declare that signes of his comming, and of other straunge effectes shall be seene in the Sonne, Moone and Starres. [...]

3.2 Edward Wright, 'Laudatory Address', in William Gilbert, *De Magnete*, 1600 (New York: Dover edition, 1958), pp. xxxvii–xlv

To the most learned Mr. William Gilbert, the distinguished London physician and father of the magnetic philosophy: a laudatory address concerning these books on magnetism, by Edward Wright.

Should there be any one, most worthy sir, who shall disparage these books and researches of yours, and who shall deem these studies trifling and in no wise sufficiently worthy of a man consecrated to the graver study of medicine, of a surety he will be esteemed no common simpleton. For that the uses of the loadstone are very considerable, yea admirable, is too well known even among men of the lowest class to call for many words from me at this time or for any commendation. In truth, in my opinion, there is no subject-matter of higher importance or of greater utility to the human race upon which you could have brought your philosophical talents to bear. For by the God-given favor of this stone has it come about that the things which for so many centuries lay hid – such vast continents of the globe, so infinite a number of countries, islands, nations and peoples – have been, almost within our own memory, easily discovered and oft explored, and that the whole circle of the globe has been circumnavigated more than once by our own Drake and Cavendish: which fact I wish to record for the undying remembrance of those men. For, by the showing of the magnetized needle, the points North, South, East and West and the other points of the compass are known to navigators, even while the sky is murky and in the deepest night; by this means seamen have understood toward what point they must steer their course, a thing that was quite impossible before the wondrous discovery of the north-pointing power of the loadstone. Hence sailors of old were often beset, as we learn from the histories, by an incredible anxiety and by great peril, for, when storms raged and the sight of sun and stars was cut off, they knew not whither they were sailing, neither could they by any means or by any device find out. Hence what must have been the gladness, what the joy of all mariners when first this magnetic pointer offered itself as a most sure guide on the route and as a God Mercury! But it was not enough for this magnetic Mercury simply to point out the way and, as it were, to show by the extended finger whither the course must be: it soon began even to indicate the distance of the place whither the voyage is made. For, since the magnetic pointer does not always regard the same northern spot in every locality, but usually varies therefrom, either to the east or to the west, tho' it nevertheless hath and holds ever the same variation in the same place, wherever that may be; it has come about that by means of this variation (as it is called) closely observed and noted in certain maritime regions, together with an observation of the latitude, the same places can afterward be found by navigators when they approach and come near to the same variation. Herein the Portuguese in their voyages to the East Indies have the surest tokens of their approaching the Cape of Good Hope, as is shown in the narrations of Hugo *Lynschetensis* and our very learned fellow-countryman Richard Hakluyt; hereby, too, many of our skilled British navigators when voyaging from the Gulf of Mexico to the Azores, can tell when they are come near to these islands, though, according to their marine charts, they may appear to be 600 English miles away. And thus, thanks to this magnetic indication, that ancient geographical problem, how to discover the longitude, would seem to be on the way to a solution; for, the variation of a seaboard place being known, that place can thereafter be very easily found as often as occasion may require, provided its latitude is not unknown.

Yet somewhat of inconvenience and difficulty seems to attach to this observation of the variation, for it cannot be made except when the sun or the stars are shining. Accordingly this magnetic Mercury of the sea, better far than Neptune himself or any of the sea gods or goddesses, proceeds still further to bestow blessings on all mariners; and not alone in the darkness of night and when the sky is murky does he show the true direction, but he seems even to give the surest indications of the latitude. For the iron pointer suspended freely and with the utmost precision in equilibrium on its axis, and then touched and excited with a loadstone, dips down to a fixed and definite point below the horizon (e.g. in the latitude of London it dips nearly 72 degrees) and there stands. But because of the wonderful agreement and congruency manifested in nearly all and singular magnetic experiments, equally in the earth itself and in a terrella (i.e. a spherical loadstone), it seems (to say the least) highly probable and more than probable that the same pointer (similarly stroked with a loadstone) will, at the equator, stand in equilibrium on the plane of the horizon. Hence, too, it is highly probable that in proceeding a very short distance from south to north (or *vice versa*) there will be a pretty sensible change in the dip; and thus the dip being carefully noted once and the latitude observed, the same place and the same latitude may thereafter be very readily found by means of a dip instrument even in the darkest night and in the thickest weather.

Thus then, to bring our discourse back again to you, most worthy and learned Mr. Gilbert (whom I gladly acknowledge as my master in this magnetical philosophy), if these books of yours on the Loadstone contained nought save this one method of finding the latitude from the magnetic dip, now first published by you, even so our British mariners as well as the French, the Dutch, the Danes, whenever they have to enter the British sea or the strait of Gibraltar from the Atlantic Ocean, will justly hold them worth no small sum of gold. And that discovery of yours, that the entire globe is magnetical, albeit to many it will seem to the last degree paradoxical, nevertheless is buttressed and confirmed by so many and so apposite experiments in Book II, Chapter XXXIV; Book III, Chapters IV and XII; and throughout nearly the whole of Book V, that no room is left for doubt or contradiction. I come therefore to the cause of magnetic variation – a problem that till now has perplexed the minds of the learned; but no one ever set forth a cause more probable than the one proposed now for the first time in these your books on the Loadstone. The fact that the magnetic needle points due north in the middle of the ocean and in the heart of continents – or at least in the heart of their more massive and more elevated parts – while near the coasts there is, afloat and ashore, an inclination of the needle toward those more massive parts, just as happens in a terrella that is made to resemble the earth globe in its greater elevation at some parts and shows that it is weak or decayed or otherwise imperfect elsewhere: all this makes exceedingly probable the theory that the variation is nothing but a deviation of the magnetic needle to those more powerful and more elevated regions of the globe. Hence the reason of the irregularity that is seen in the variations of the compass is easily found in the inequality and anomaly of those more elevated parts. Nor do I doubt that all those who have imagined or accepted certain 'respective points' as well as they who speak of magnetic mountains or rocks or poles, will begin to waver as soon as they read these your books on the Loadstone and will of their own accord come over to your opinion.

As for what you have finally to say of the circular motion of the earth and the terrestrial poles, though many will deem it the merest theorizing, still I do not see why it should not meet with indulgence even among those who do not acknowledge the earth's motion to be spherical, seeing that even they cannot readily extricate themselves from the many difficulties that result from a diurnal motion of the whole heavens. For, first, it is not reasonable to have that done by many agents which can be done by fewer, or to have the whole heavens and all the spheres (if spheres there be) of the planets and fixed stars made to revolve for the sake of the diurnal motion, which may be accounted for by a daily rotation of the earth. Then, which theory is the more probable, that the equinoctial circle of the earth may make a rotatary movement of one quarter of an English mile (60 miles being equal to one degree on the earth's equator) in one second of time, i.e., in about as much time as it takes to make only one step when one is walking rapidly; or that the equator of the *primum mobile* in the same time, with inexpressible celerity, makes 5000 miles and that in the twinkling of an eye it makes about 50 English miles, surpassing the velocity of a flash of lightning, if they are in the right who most strenuously deny the earth's motion? Finally, which is the more probable, to suppose that this little globe of the earth has some motion, or with mad license of conjecture to superpose three mighty starless spheres, a ninth, a tenth, and an eleventh, upon the eighth sphere of the fixed stars, particularly when from these books on the Loadstone and the comparison of the earth with the terrella it is plain that spherical motion is not so contrary to the nature of the earth as it is commonly supposed to be?

Nor do the passages quoted from Holy Writ appear to contradict very strongly the doctrine of the earth's mobility. It does not seem to have been the intention of Moses or the prophets to promulgate nice mathematical or physical distinctions: they rather adapt themselves to the understanding of the common people and to the current fashion of speech, as nurses do in dealing with babes; they do not attend to unessential minutiæ. Thus, Genesis i. 16 and Psalm cxxxvi. 7, 9, the moon is called a great luminary, because it so appears to us, though, to those versed in astronomy, it is known that very many stars, fixed and planetary, are far larger. So, too, from Ps. civ. 5, no argument of any weight can, I think, be drawn to contradict the earth's mobility, albeit it is said that God established the earth on her foundations to the end it should never be moved; for the earth may remain forevermore in its own place and in the selfsame place, in such manner that it shall not be moved away by any stray force of transference, nor carried beyond its abiding place wherein it was established in the beginning by the divine architect. We, therefore, while we devoutly acknowledge and adore the inscrutable wisdom of the triune Godhead, having with all diligence investigated and discerned the wondrous work of his hands in the magnetic movements, do hold it to be entirely probable, on the ground of experiments and philosophical reasons not few, that the earth while it rests on its centre as its basis and foundation, hath a spherical motion nevertheless.

But, apart from these matters (touching which no one, I do believe, ever gave more certain demonstrations), no doubt your discussion of the causes of variation and of the dip of the needle beneath the horizon (to say nothing of sundry other points which 'twould take too long to mention) will find the heartiest approval among all intelligent men and 'children, of magnetic science' (to use the language of the chemists). Nor have I any doubt that, by publishing these your books on the Loadstone, you will stimulate all wide-awake navigators to give not less study to observation of dip than of variation. For it is highly probable, if not certain, that latitude, or rather the effect of latitude, can be determined much more accurately (even when the sky is darkest) from the dip alone, than longitude or the effect of longitude can be found from the variation even in the full light of day or while all the stars are shining, and with the help of the most skilfully and ingeniously contrived instrument. Nor is there any doubt that those most learned men, Petrus Plantius (a most diligent student not so much of geography as of magnetic observations) and Simon Stevinius, a most eminent mathematician, will be not a little rejoiced when first they set eyes on these your books and therein see their own *λιμνευρετικήν* or method of finding ports so greatly and unexpectedly enlarged and developed; and of course they will, as far as they may be able, induce all navigators among their own countrymen to note the dip no less than the variation of the needle.

Let your magnetic Philosophy, most learned Mr. Gilbert, go forth then under the best auspices – that work held back not for nine years only, according to Horace's Counsel, but for almost other nine; that Philosophy which by your multitudinous labors, studies, vigils, and by your skill and at your no inconsiderable expense has been after long years at last, by means of countless ingenious experiments, taken bodily out of the darkness and dense murkiness with which it was surrounded by the speculations of incompetent and shallow philosophizers; nor did you in the mean time overlook, but did diligently read and digest whatever had been published in the writings whether of the ancients or the moderns. [...]

3.3 Tycho Brahe, *De disciplinis mathematicis oratio*, 1574, trans. and ed. P. Maxwell-Stuart in *The Occult in Early Modern Europe* (Basingstoke: Macmillan, 1999), pp. 84–5

To deny the power and influence of the stars is to detract from divine wisdom and influence. What more prejudiced or what sillier thought could one have about God than that He had made the most enormous and extraordinary of all heavens and a theatre of so many shining stars in vain and to no useful purpose, when no human being does even the most worthless task except for a particular purpose?...We see a great diversity of natures in individual human beings. Some who have been shaped under the fortunate influence of Saturn, the highest star, investigate in solitude matters which are exalted and far beyond the understanding of common people. There are those who have a greater interest in judicial and political affairs, and upon them the brilliance of Jupiter has looked with favour. There are several – those who are stirred up by the hot passion of Mars – who do not breathe at all unless it be for wars, slaughters, disorders and altercations. Others, under the ambitious influence of the Sun, strive after honours and dignities, and seek to control things; and there are those under the spell of Venus, the seductive star, who spend their lives in love-affairs, giving and receiving pleasure, music, and other delicious delights. Others, aroused by Mercury, devote themselves completely to the exercise of other remarkable talents, or even to commerce. Some, allocated the influence of a lunar nature, spend the whole of their lives doing ordinary things, foreign travel, sea-voyages, fishing and stuff like that. In this way one can see that the great variety of dispositions mirrors the influences of the seven planets. However, most people are affected by various combinations of these planets and they pursue different types of activity at different stages of their lives, now occupied with one kind of business, now with another, according to the way they are subjected to a greater or lesser extent at one time or another to the rays of a planet by means of hidden progressions. These differences may be seen in brothers born to the same parents in the same place and brought up in similar fashion, to such an extent that brothers very often differ in nature and character more than any other people. The reasons for these and similar diversities can be sought nowhere better than in astrology, although I do not deny that people can also be changed by lesser causes such as upbringing, education, conversation, foreign travel and things like that. ...Philosophers maintain that astrology should not be included among the other arts because it lacks reliable, clear principles. It is impossible, they say, to know the exact moment of birth and hence astrologers vainly fall back on the moment of conception, since this is a lot less uncertain. They adduce further arguments hostile (as they see it) to the art – that many people are born at the same time and yet have different fates and different things happen to them during their lives. Twins born of the same parents at almost the same time are very often allotted completely different fortunes; very many die at one and the same time, in war, in time of plague, during other general catastrophes, and yet their horoscope could not have signified the same type of death at all....Theologians agree with these arguments and add others, proclaiming that this art has been forbidden by the word of God and that it impiously leads people away from the knowledge of God and acts against good morals....But those who say that people born at the same time have different fortunes know that astrologers do not claim that the sky acts in precisely the same way upon all those born at one and the same time, but that they are subject to diversity and are altered in different ways by heavenly influences. Very often the influence of the stars is shifted about in different ways for different reasons, such as upbringing, schooling, conversation and similar changing circumstances in life. Nor is man's free will in any way made subordinate to the stars but through it, under the guidance of reason, he can do very many things beyond the influence of the stars, if that is what he wishes. Astrologers do not require everyone to receive the influences of the stars in the same way, but some more and some less, according to their aptitude for receiving them or their immunity to them.... Theologians do not take into consideration that astrologers do not bind people's will to the stars, but agree that there is something in humanity which has been raised above all the stars and whose beneficial effect is that if people wish to live as true and supra-mundane human beings, they can conquer whatever malevolent inclinations they may have from the stars. But if they choose to live a brutish life, to be carried along by blind emotions, and prefer to fornicate with the beasts, they must not think God is the author of this mistake; for God created human beings in such a way that they can, if they wish, overcome all malevolent inclinations they get from the stars.

3.4 J. Kepler, *Astronomia Nova*, Heidelberg, 1609, trans. C. A. Russell, in D. C. Goodman (ed.), *Science and Religious Belief, 1600–1900: A Selection of Primary Sources* (John Wright and Open University Press, 1973), pp. 22–3

See Chapter 10 of my *Astronomy: the Optical Section* where you will find reasons why the sun in this way seems to all men to be moving, but not the earth: namely, because the sun seems small, but the earth truly appears to be large. Nor is the motion of the sun to be grasped by sight (since it gives the appearance of being slow) but by reason alone on account of the changed relationship to the mountains after some time. It is therefore impossible that reason not previously instructed should imagine anything other than that the earth is a kind of vast house with the vault of the sky placed on top of it; it is motionless and within it the sun being so small passes from one region to another, like a bird wandering through the air.

This universal image has produced the first line in the sacred page. *In the beginning*, said Moses, *God created the heaven and the earth*; this is a natural expression because these two aspects of the universe are those that chiefly meet the eye. It is as if Moses were saying to man 'all this architecture of the universe that you see, the brightness above, by which you are covered, the widespreading darkness below, upon which you stand – all this had been created by God'.

In other places man is questioned whether he has learned how to penetrate the height of the sky above or the depth of the earth beneath. This is natural because to the mass of men each of these appears equally to project into infinite space. Nevertheless, there never was a man who, listening rationally, would use these words to circumscribe the diligence of the astronomers, whether in demonstrating the most contemptible weakness of the earth by comparison with the sky, or through investigations of astronomical distance. These words do not speak about intellectualised dimensions, but about the dimension of reality – which, for a human body fixed on the earth and drinking in the free air, is totally impossible. Read the whole of Job Ch. 38 and compare with it the matters that are disputed in astronomy and physics.

If anyone alleges on the basis of Psalm 24 *The earth is founded upon the seas* (in order to establish some new philosophical dictum, however absurd to hear) that the earth is floating on the waters, may it not be rightly said to him that he ought to set free the Holy Spirit and should not drag Him in to the schools of physics to make a fool of Him. For in that place the Psalmist wishes to suggest nothing other than what men know beforehand and experience each day: the lands, uplifted after separation of the waters, have great rivers flowing through them and the seas around them on all sides. Doubtless the same is spoken of elsewhere, when the Israelites sing *By the waters of Babylon there we sat down*, i.e., by the side of the rivers, or on the banks of the Euphrates and Tigris.

If anyone receives the one freely, why not the other, so that in other places which are often quoted against the motion of the earth we should, in the same way, turn our eyes from physics to the tradition of scripture?

One generation passes away, says Ecclesiastes, *and another generation is born*, but the earth abides for ever. Is Solomon here, as it were, disputing with the astronomers? No, he is rather warning men of their changeableness whereas the earth, the home of the human race, always remains the same; the movement of the sun keeps returning it to its starting-point; the wind is driven in a circle, and returns to the same plan; rivers flow from their sources to the sea, and thence return to their sources. Finally, while some men perish others are born, and always the drama of life is the same; there is nothing new under the sun.

You are listening to no new principle of physics. It is a question of ethical instruction in a matter which is clear on its own, observed universally but receives scant consideration. That is why Solomon insists on the matter. Who does not know the earth to be always the same? Who does not see that the sun rising daily in the East, that the rivers run perpetually down to the sea, that the pattern of changes of the wind is fixed and recurring and that one generation succeeds another? Who in fact considers that the drama of life is being perpetually performed, with only a change of cast and that there is nothing new in human affairs? And so, by rehearsing things which everyone sees, Solomon warns of that which the majority wrongly neglect.

But some men think Psalm 104 to be wholly concerned with physics, since it is wholly concerned with physical matters. And there God is said to have *laid the foundations of the earth so that it should not be moved*, and that stability will remain from age to age. Nevertheless the Psalmist is a very long way from speculation about physical causes. He rests utterly in the greatness of God who made all these things and is unfolding a hymn to God the Creator, a hymn in which he runs in order through the whole world as it appears to our eyes.

3.5 Johannes Kepler, *Harmonices Mundi* (*The Harmonies of the World*), 1619, trans. C. G. Wallis, in *Great Books of the World*, vol. 16 (Chicago: Encyclopaedia Britannica Inc., 1952), pp. 1014–18, 1040–1

3. A Summary of Astronomical Doctrine Necessary for Speculation into the Celestial Harmonies

First of all, my readers should know that the ancient astronomical hypotheses of Ptolemy, in the fashion in which they have been unfolded in the *Theoricae* of Peurbach and by the other writers of epitomes, are to be completely removed from this discussion and cast out of the mind. For they do not convey the true lay out of the bodies of the world and the polity of the movements.

Although I cannot do otherwise than to put solely Copernicus' opinion concerning the world in the place of those hypotheses and, if that were possible, to persuade everyone of it; but because the thing is still new among the mass of the intelligentsia [*apud vulgus studiosorum*], and the doctrine that the Earth is one of the planets and moves among the stars around a motionless sun sounds very absurd to the ears of most of them: therefore those who are shocked by the unfamiliarity of this opinion should know that these harmonical speculations are possible even with the hypotheses of Tycho Brahe – because that author holds, in common with Copernicus, everything else which pertains to the lay out of the bodies and the tempering of the movements, and transfers solely the Copernican annual movement of the Earth to the whole system of planetary spheres and to the sun, which occupies the centre of that system, in the opinion of both authors. For after this transference of movement it is nevertheless true that in Brahe the Earth occupies at any time the same place that Copernicus gives it, if not in the very vast and measureless region of the fixed stars, at least in the system of the planetary world. And accordingly, just as he who draws a circle on paper makes the writing-foot of the compass revolve, while he who fastens the paper or tablet to a turning lathe draws the same circle on the revolving tablet with the foot of the compass or stylus motionless; so too, in the case of Copernicus the Earth, by the real movement of its body, measures out a circle revolving midway between the circle of Mars on the outside and that of Venus on the inside; but in the case of Tycho Brahe the whole planetary system (wherein among the rest the circles of Mars and Venus are found) revolves like a tablet on a lathe and applies to the motionless Earth, or to the stylus on the lathe, the midspace between the circles of Mars and Venus; and it comes about from this movement of the system that the Earth within it, although remaining motionless, marks out the same circle around the sun and midway between Mars and Venus, which in Copernicus it marks out by the real movement of its body while the system is at rest. Therefore, since harmonic speculation considers the eccentric movements of the planets, as if seen from the sun, you may easily understand that if any observer were stationed on a sun as much in motion as you please, nevertheless for him the Earth, although at rest (as a concession to Brahe), would seem to describe the annual circle midway between the planets and in an intermediate length of time. Wherefore, if there is any man of such feeble wit that he cannot grasp the movement of the earth among the stars, nevertheless he can take pleasure in the most excellent spectacle of this most divine construction, if he applies to their image in the sun whatever he hears concerning the daily movements of the Earth in its eccentric – such an image as Tycho Brahe exhibits, with the Earth at rest.

And nevertheless the followers of the true Samian philosophy have no just cause to be jealous of sharing this delightful speculation with such persons, because their joy will be in many ways more perfect, as due to the consummate perfection of speculation, if they have accepted the immobility of the sun and the movement of the earth.

Firstly [I], therefore, let my readers grasp that today it is absolutely certain among all astronomers that all the planets revolve around the sun, with the exception of the moon, which alone has the Earth as its centre: the magnitude of the moon's sphere or orbit is not great enough for it to be delineated in this diagram in a just ratio to the rest. Therefore, to the other five planets, a sixth, the Earth, is added, which traces a sixth circle around the sun, whether by its own proper movement with the sun at rest, or motionless itself and with the whole planetary system revolving.

Secondly [II]: It is also certain that all the planets are eccentric, *i.e*, they change their distances from the sun, in such fashion that in one part of their circle they become farthest away from the sun, and in the opposite part they come nearest to the sun. In the accompanying diagram three circles apiece have been drawn for the single planets: none of them indicate the eccentric route of the planet itself; but the mean circle, such as *BE* in the case of Mars, is equal to the eccentric orbit, with respect to its longer diameter. But the orbit itself, such as *AD*, touches *AF*, the upper of the three, in one place *A*, and the lower circle *CD*, in the opposite place *D*. The circle *GH* made with dots and described through the centre of the sun indicates the route of the sun according to Tycho Brahe. And if the sun moves on this route, then absolutely all the points in this whole planetary system here depicted advance upon an equal route, each upon his own. And with one point of it (namely, the centre of the sun) stationed at one point of its circle, as here at the lowest, absolutely each and every point of the system will be stationed at the lowest part of its circle. However, on account of the

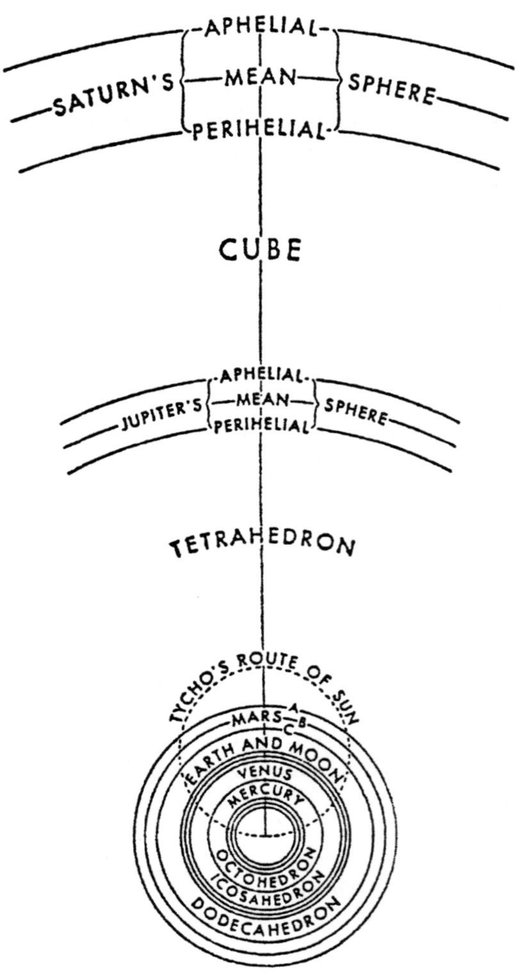

smallness of the space the three circles of Venus unite in one, contrary to my intention.

Thirdly [III]: Let the reader recall from my *Mysterium Cosmographicum*, which I published twenty-two years ago, that the number of the planets or circular routes around the sun was taken by the very wise Founder from the five regular solids, concerning which Euclid, so many ages ago, wrote his book which is called the *Elements* in that it is built up out of a series of propositions. But it has been made clear in the second book of this work that there cannot be more regular bodies, *i.e.*, that regular plane figures cannot fit together in a solid more than five times.

Fourthly [IV]: As regards the ratio of the planetary orbits, the ratio between two neighbouring planetary orbits is always of such a magnitude that it is easily apparent that each and every one of them approaches the single ratio of the spheres of one of the five regular solids, namely, that of the sphere circumscribing to the sphere inscribed in the figure. Nevertheless it is not wholly equal, as I once dared to promise concerning the final perfection of astronomy. For, after completing the demonstration of the intervals from Brahe's observations, I discovered the following: if the angles of the cube are applied to the inmost circle of Saturn, the centres of the planes are approximately tangent to the middle circle of Jupiter; and if the angles of the tetrahedron are placed against the inmost circle of Jupiter, the centres of the planes of the tetrahedron are approximately tangent to the outmost circle of Mars; thus if the angles of the octahedron are placed against any circle of Venus (for the total interval between the three has been very much reduced), the centres of the planes of the octahedron penetrate and descend deeply within the outmost circle of Mercury, but nonetheless do not reach as far as the middle circle of Mercury; and finally, closest of all to the ratios of the dodecahedral and icosahedral spheres – which ratios are equal to one another – are the ratios or intervals between the circles of Mars and the Earth, and the Earth and Venus; and those intervals are similarly equal, if we compute from the inmost circle of Mars to the middle circle of the Earth, but from the middle circle of the Earth to the middle circle of Venus. For the middle distance of the Earth is a mean proportional between the least distance of Mars and the middle distance of Venus. However, these two ratios between the planetary circles are still greater than the ratios of those two pairs of spheres in the figures, in such fashion that the centres of the dodecahedral planes are not tangent to the outmost circle of the Earth, and the centres of the icosahedral planes are not tangent to the outmost circle of Venus; nor, however, can this gap be filled by the semidiameter of the lunar sphere, by adding it, on the upper side, to the greatest distance of the Earth and subtracting it, on the lower, from the least distance of the same. But I find a certain other ratio of figures – namely, if I take the augmented dodecahedron, to which I have given the name of echinus, (as being fashioned from twelve quinquangular stars and thereby very close to the five regular solids), if I take it, I say, and place its twelve points in the inmost circle of Mars, then the sides of the pentagons, which are the bases of the single rays or points, touch the middle circle of Venus. In short: the cube and the octahedron, which are consorts, do not penetrate their planetary spheres at all; the dodecahedron and the icosahedron, which are consorts, do not wholly reach to theirs, the tetrahedron exactly touches both: in the first case there is falling short; in the second, excess; and in the third, equality, with respect to the planetary intervals.

Wherefore it is clear that the very ratios of the planetary intervals from the sun have not been taken from the regular solids alone. For the Creator, who is the very source of geometry and, as Plato wrote, 'practices eternal geometry', does not stray from his own archetype. And indeed that very thing could be inferred from the fact that all the planets change their intervals throughout fixed periods of time, in such fashion that each has two marked intervals from the sun, a greatest and a least; and a fourfold comparison of the intervals from the sun is possible between two planets: the comparison can be made between either the greatest, or the least, or the contrary intervals most remote from one another, or the contrary intervals nearest together. In this way the comparisons made two by two between neighbouring planets are twenty in number, although on the contrary there are only five regular solids. But it is consonant that if the Creator had any concern for the ratio of the spheres in general, He would also have had concern for the ratio which exists between the varying intervals of the single planets specifically and that the concern is the same in both cases and the one is bound up with the other. If we ponder that, we will comprehend that for setting up the diameters and eccentricities conjointly, there is need of more principles, outside of the five regular solids. [...]

7. The Universal Consonances of All Six Planets, Like Common Four-Part Counterpoint, Can Exist

But now, Urania, there is need for louder sound while I climb along the harmonic scale of the celestial movements to higher things where the true archetype of the fabric of the world is kept hidden. Follow after, ye modern musicians, and judge the thing according to your arts, which were unknown to antiquity. Nature, which is never not lavish of herself, after a lying-in of two thousand years, has finally brought you forth in these last generations, the first true images of the universe. By means of your concords of various voices, and through your ears, she has whispered to the human mind, the favorite daughter of God the Creator, how she exists in the innermost bosom.

(Shall I have committed a crime if I ask the single composers of this generation for some artistic motet instead of this epigraph? The Royal Psalter and the other Holy Books can supply a text suited for this. But alas for you! No more than six are in concord in the heavens. For the moon sings her monody separately, like a dog sitting on the Earth. Compose the melody; I, in order that the book may progress, promise that I will watch carefully over the six parts. To him who more properly expresses the celestial music described in this work, Clio will give a garland, and Urania will betroth Venus his bride.)

It has been unfolded above what harmonic ratios two neighbouring planets would embrace in their extreme movements. But it happens very rarely that two, especially the slowest, arrive at their extreme intervals at the same time; For example, the apsides of Saturn and Jupiter are about 81° apart. Accordingly, while this distance between them measures out the whole zodiac by definite twenty-year leaps, eight hundred years pass by, and nonetheless the leap which concludes the eighth century, does not carry precisely to the very apsides; and if it digresses much further, another eight hundred years must be awaited, that a more fortunate leap than that one may be sought; and the whole route must be repeated as many times as the measure of digression is contained in the length of one leap. Moreover, the other single pairs of planets have periods like that, although not so long. But meanwhile there occur also other consonances of two planets, between movements whereof not both are extremes but one or both are intermediate; and those consonances exist as it were in different tunings [*tensionibus*]. For, because Saturn tends from G to *b*, and slightly further, and Jupiter from *b* to *d* and further; therefore between Jupiter and Saturn there can exist the following consonances, over and above the octave: the major and minor third and the perfect fourth, either one of the thirds through the tuning which maintains the amplitude of the remaining one, but the perfect fourth through the amplitude of a major whole tone. For there will be a perfect fourth not merely from G of Saturn to *cc* of Jupiter but also from *A* of Saturn to *dd* of Jupiter and through all the intermediates between the *G* and *A* of Saturn and the *cc* and *dd* of Jupiter. But the octave and the perfect fifth exist solely at the points of the apsides. But Mars, which got a greater interval as its own, received it in order that it should also make an octave with the upper planets through some amplitude of tuning. Mercury received an interval great enough for it to set up almost all the consonances with all the planets within one of its periods, which is not longer than the space of three months. On the other hand, the Earth, and Venus much more so, on account of the smallness of their intervals, limit the consonances, which they form not merely with the others but with one another in especial, to visible fewness. But if three planets are to concord in one harmony, many periodic returns are to be awaited; nevertheless there are many consonances, so that they may so much the more easily take place, while each nearest consonance follows after its neighbour, and very often threefold consonances are seen to exist between Mars, the Earth, and Mercury. But the consonances of four planets now begin to be scattered throughout centuries, and those of five planets throughout thousands of years.

But that all six should be in concord has been fenced about by the longest intervals of time; and I do not know whether it is absolutely impossible for this to occur twice by precise evolving or whether that points to a certain beginning of time, from which every age of the world has flowed.

But if only one sextuple harmony can occur, or only one notable one among many, indubitably that could be taken as a sign of the Creation. Therefore we must ask, in exactly how many forms are the movements of all six planets reduced to one common harmony? ...

3.6 John Wilkins, *A Discourse Concerning a New Planet*, 1640, in *The Mathematical and Philosophical Works of the Right Rev. John Wilkins*, 1802 (London: Frank Cass and Co. Ltd, 1970), pp. 138–46

[. . .] In weighing the authority of others, it is not their multitude that should prevail, or their skill in some things that should make them of credit in every thing; but we should examine what particular insight and experience they had in those things for which they are cited. Now it is plain, that common people judge by their senses, and therefore their voices are altogether unfit to decide any philosophical doubt, which cannot well be examined or explained without discourse and reason. And as for the ancient fathers, though they were men very eminent for their holy lives, and extraordinary skill in divinity, yet they were most of them very ignorant in that part of learning which concerns this opinion; as appears by many of their gross mistakes in this kind; as that concerning the antipodes, &c. and therefore it is not their opinion neither, in this business, that to an indifferent seeker of truth will be of any strong authority.

But against this it is objected*. That the instance of the antipodes does not argue any special ignorance in these learned men; or that they had less skill in such human arts than others; since Aristotle himself, and Pliny, did deny this as well as they.

I answer:

1. If they did, yet this does make more to the present purpose: for if such great scholars, who were so eminent for their knowledge in natural things, might yet notwithstanding be grossly mistaken in such matters as are now evident and certain, why then we have no reason to depend upon their assertions or authorities, as if they were infallible.

2. Though these great naturalists, for want of some experience, were mistaken in that opinion, whilst they thought no place was habitable but the temperate zones: yet it cannot be from hence inferred that they denied the possibility of antipodes; since these are such inhabitants as live opposite unto us in the other temperate zone: and it were an absurd thing to imagine that those who lived in different zones, can be antipodes to one another; and argues that a man did not understand, or else had forgotten that common distinction in geography, wherein the relation of the world's inhabitants unto one another are reckoned up under these three heads; *antaci, periæci,* and *antipodes*. But to let this pass: it is certain, that some of the fathers did deny the being of any such, upon other more absurd grounds. Now if such as Chrysostom, Lactantius, &c. who were noted for great scholars; and such too as flourished in these latter times, when all human learning was more generally professed, should notwithstanding be so much mistaken in so obvious a matter: why then may we not think that those primitive saints, who were the penmen of scripture, and eminent above others in their time

* Alex. Ross. l. 1. sect. c. 8.

for holiness and knowledge; might yet be utterly ignorant of many philosophical truths, which are commonly known in these days? It is probable, that the Holy Ghost did inform them only with the knowledge of those things whereof they were to be the penmen, and that they were not better skilled in points of philosophy than others. There were indeed some of them who were supernaturally endowed with human learning; yet this was, because they might thereby be fitted for some particular ends, which all the rest were not appointed unto: thus Solomon was strangely gifted with all kind of knowledge, in a great measure; because he was to teach us by his own experience the extreme vanity of it, that we might not so settle our desires upon it, as if it were able to yield us contentment*. So too the apostles were extraordinarily inspired with the knowledge of languages, because they were to preach unto all nations. But it will not hence follow, that therefore the other holy penmen were greater scholars than others. It is likely that Job had as much human learning as most of them, because his book is more especially remarkable for lofty expressions, and discourses of nature; and yet it is not likely that he was acquainted with all those mysteries which later ages have discovered; because when God would convince him of his own folly and ignorance, he proposes to him such questions, as being altogether unanswerable; which notwithstanding, any ordinary philosopher in these days might have resolved. [...]

3. It is considerable, that in the rudiments and first beginnings of astronomy, and so in several ages after, this opinion hath found many patrons, and those too men of eminent note and learning. Such was more especially Pythagoras, who was generally and highly esteemed for his divine wit, and rare inventions; under whose mysterious sayings, there be many excellent truths to be discovered.

But against his testimony, it is again objected††; if Pythagoras were of this opinion, yet his authority should not be of any credit, because he was the author of many other monstrous absurdities.

To this I answer; if a man's error in some particulars should take away his credit for every thing else, this would abolish the force of all human authority; for *humanum est errare*. Secondly, it is probable that many of Pythagoras's sayings which seem so absurd, are not to be understood according to their letter, but in a mystical sense. [...]

4. It is considerable, that since this science of astronomy hath been raised to any perfection, there have been many of the best skill in it, that have assented unto that assertion which is here defended. Amongst whom was the cardinal Cusanus†, but more especially Copernicus, who was a man very exact and diligent in these studies for above thirty years together, from the year 1500 to 1530, and upwards; and since him, most of the best

---

* Eccl. i. 18.

†† Alex. Ros. l. 2. sect. 2. c. 10.

† De doct. ignor. l. 2. cap. 12.

astronomers have been of this side. So that now there is scarce any of note and skill, who are not Copernicus's followers; and if we should go to most voices, this opinion would carry it from any other. It would be too tedious to reckon up the names of those that may be cited for it; I will only mention some of the chief; such were Joachinus Rheticus, an elegant writer; Christopherus Rothman; Mestlin, a man very eminent for his singular skill in this science; who though at the first he were a follower of Ptolemy, yet upon his second and more exact thoughts, he concluded Copernicus to be in the right; and that the usual hypothesis, *præscriptione potius quam ratione valet**, does prevail more by prescription than reason. So likewise Erasmus Reinoldus, who was the man that calculated the prutenical tables from Copernicus his Observations, and did intend to write a commentary upon his other works, but that he was taken out of this life before he could finish those resolutions. Unto these also I might add the names of Gilbert, Keplar, Galilæus, with sundry others, who have much beautified and confirmed this hypothesis, with their new inventions[†]. Nay I may safely affirm, that amongst the variety of those opinions that are in astronomy, there are more (of those which have skill in it) that are of this opinion, not only than any other side, but than all the rest put together. So that now it is a greater argument of singularity to oppose it.

5. It is probable, that many other of the ancients would have assented unto this opinion, if they had been acquainted with those experiments which later times have found out for the confirmation of it: and therefore Rheticus[‡] and Keplar[||] do so often wish that Aristotle were now alive again. Questionless, he was so rational and ingenious a man, (not half so obstinate as many of his followers) that upon such probabilities as these, he would quickly have renounced his own principles, and have come over to this side: for in one place, having proposed some questions about the heavens[§], which were not easy to be resolved, he sets down this rule; that in difficulties, a man may take a liberty to speak that which seems most likely to him; and in such cases, an aptness to guess at some resolution, for the satisfying of our philosophical thirst, does deserve rather to be stiled by the name of modesty, than boldness. And in another place[¶], he refers the reader to the different opinions of astronomers, advising him to examine their several tenets, as well Eudoxus as Calippus; and to entertain that (not which is most ancient, but) which is most exact and agreeable to reason. And as for Ptolomy, it is his counsel[**], that we should endeavour to frame such

* Præf. ad Narrat. Rhetici.

† Ibid.

‡ In narratione.

|| Myst. Cosmogr. c. 1. item Præf. ad 4. 1. Astr. Copern.

§ De Cœl. l. 2. c. 12.

¶ Met. lib. 12. cap. 8.

** Alm. l. 13. cap. 2.

suppositions of the heavens, as might be more simple, being void of all superfluities: and he confesses, that his hypothesis had many implications in it, together with sundry intricate and unlikely turnings; and therefore in the same place, he seems to admonish us, that we should not be too confident the heavens were really in the same form wherein astronomers did suppose them. So that it is likely, it was his chief intent to propose unto us such a frame of the celestial bodies, from which we might, in some measure, conceive of their different appearances; and according to which, we might be able to calculate their motions. But now it is Copernicus's endeavour, to propound unto us the true natural causes of these several motions and appearances: it was the intent of the one, to settle the imagination; and of the other, to satisfy the judgment. So that we have no reason to doubt of his assent unto this opinion, if he had but clearly understood all the grounds of it.

...Whereas on the contrary, there are very few to be found amongst the followers of Aristotle and Ptolomy, that have read any thing in Copernicus, or do fully understand the grounds of his opinion; and I think, not any, who having been once settled with any strong assent on this side, that have afterwards revolted from it. Now if we do but seriously weigh with ourselves, that so many ingenious, considering men, should reject that opinion which they were nursed up in, and which is generally approved as the truth; and that for the embracing of such a paradox as is condemned in schools, and commonly cried down, as being absurd and ridiculous; I say, if a man do but well consider all this, he must needs conclude, that there is some strong evidence for it to be found out by examination; and that in all probability, this is the righter side. [...]

Chapter Four
Crisis in Italy

4.1 (a) Galileo, *Letter to the Grand Duchess Christina* [1615], 1636, and (b) Cardinal Bellarmine, letter to Paolo Foscarini, 12 April 1615, both trans. Stillman Drake in *Discoveries and Opinions of Galileo* (New York: Doubleday, 1957), pp. 181–200, 162–4

(a) Galileo, Letter to the Grand Duchess Christina

The reason produced for condemning the opinion that the earth moves and the sun stands still is that in many places in the Bible one may read that the sun moves and the earth stands still. Since the Bible cannot err, it follows as a necessary consequence that anyone takes an erroneous and heretical position who maintains that the sun is inherently motionless and the earth movable.

With regard to this argument, I think in the first place that it is very pious to say and prudent to affirm that the holy Bible can never speak untruth – whenever its true meaning is understood. But I believe nobody will deny that it is often very abstruse, and may say things which are quite different from what its bare words signify. Hence in expounding the Bible if one were always to confine oneself to the unadorned grammatical meaning, one might fall into error. Not only contradictions and propositions far from true might thus be made to appear in the Bible, but even grave heresies and follies. Thus it would be necessary to assign to God feet, hands, and eyes, as well as corporeal and human affections, such as anger, repentance, hatred, and sometimes even the forgetting of things past and ignorance of those to come. These propositions uttered by the Holy Ghost were set down in that manner by the sacred scribes in order to accommodate them to the capacities of the common people, who are rude and unlearned. For the sake of those who deserve to be separated from the herd, it is necessary that wise expositors should produce the true senses of such passages, together with the special reasons for which they were set down in these words. This doctrine is so

THE
Ancient and Modern
DOCTRINE
OF
Holy Fathers,
AND
Iudicious Divines,
CONCERNING
The rash citation of the Testimony of SACRED SCRIPTURE, in Conclusions meerly Natural, and that may be proved by Sensible Experiments, and Necessary Demonstrations.

Written, some years since, to Gratifie The most SERENE CHRISTINA LOTHARINGA, *Arch-Dutchess* of *TUSCANY*;

By GALILÆO GALILÆI, A Gentleman of *Florence*, and Chief Philosopher and Mathematician to His most Serene Highness the Grand *DUKE*.

And now rendred into English from the Italian,
BY
THOMAS SALUSBURY.

Naturam Rerum invenire, difficile; & ubi inveneris, indicare in vulgus, nefas. Plato.

LONDON,
Printed by WILLIAM LEYBOURN, 1661.
Hhh

The title-page of Galileo's Letter to the Grand Duchess *in Thomas Salusbury's translation,* Mathematical Collections *(London, 1661).*

widespread and so definite with all theologians that it would be superfluous to adduce evidence for it.

Hence I think that I may reasonably conclude that whenever the Bible has occasion to speak of any physical conclusion (especially those which are very abtruse and hard to understand), the rule has been observed of avoiding confusion in the minds of the common people which would render them contumacious toward the higher mysteries. Now the Bible, merely to condescend to popular capacity, has not hesitated to obscure some very important pronouncements, attributing to God himself some qualities extremely remote from (and even contrary to) His essence. Who, then, would positively declare that this principle has been set aside, and the Bible has confined itself rigorously to the bare and restricted sense of its words, when speaking but casually of the earth, of water, of the sun, or of any other created thing? Especially in view of the fact that these things in no way concern the primary purpose of the sacred writings, which is the service of God and the salvation of souls – matters infinitely beyond the comprehension of the common people.

This being granted, I think that in discussions of physical problems we ought to begin not from the authority of scriptural passages, but from sense-experiences and necessary demonstrations; for the holy Bible and the phenomena of nature proceed alike from the divine Word, the former as the dictate of the Holy Ghost and the latter as the observant executrix of God's commands. It is necessary for the Bible, in order to be accommodated to the understanding of every man, to speak many things which appear to differ from the absolute truth so far as the bare meaning of the words is concerned. But Nature, on the other hand, is inexorable and immutable; she never transgresses the laws imposed upon her, or cares a whit whether her abstruse reasons and methods of operation are understandable to men. For that reason it appears that nothing physical which sense-experience sets before our eyes, or which necessary demonstrations prove to us, ought to be called in question (much less condemned) upon the testimony of biblical passages which may have some different meaning beneath their words. For the Bible is not chained in every expression to conditions as strict as those which govern all physical effects; nor is God any less excellently revealed in Nature's actions than in the sacred statements of the Bible. [...]

But I do not feel obliged to believe that that same God who has endowed us with senses, reason, and intellect has intended to forgo their use and by some other means to give us knowledge which we can attain by them. He would not require us to deny sense and reason in physical matters which are set before our eyes and minds by direct experience or necessary demonstrations. This must be especially true in those sciences of which but the faintest trace (and that consisting of conclusions) is to be found in the Bible. Of astronomy, for instance, so little is found that none of the planets except Venus are so much as mentioned, and this only once or twice under the name of 'Lucifer'. If the sacred scribes had had any intention of teaching people certain arrangements and motions of the heavenly bodies, or had they wished us to derive such knowledge from the Bible, then in my opinion they would not have spoken of these matters so sparingly in comparison with the infinite number of admirable conclusions which are demonstrated in that science. Far from pretending to teach us the constitution and motions of the heavens and the stars, with their shapes, magnitudes, and distances, the authors of the Bible intentionally forbore to speak of these things, though all were quite well known to them. [...]

Let us grant then that theology is conversant with the loftiest divine contemplation, and occupies the regal throne among sciences by dignity. But acquiring the highest authority in this way, if she does not descend to the lower and humbler speculations of the subordinate sciences and has no regard for them because they are not concerned with blessedness, then her professors should not arrogate to themselves the authority to decide on controversies in professions which they have neither studied nor practiced. Why, this would be as if an absolute despot, being neither a physician nor an architect but knowing himself free to command, should undertake to administer medicines and erect buildings according to his whim – at grave peril of his poor patients' lives, and the speedy collapse of his edifices.

Again, to command that the very professors of astronomy themselves see to the refutation of their own observations and proofs as mere fallacies and sophisms is to enjoin something that lies beyond any possibility of accomplishment. For this would amount to commanding that they must not see what they see and must not understand what they know, and that in searching they must find the opposite of what they actually encounter. Before this could be done they would have to be taught how to make one mental faculty command another, and the inferior powers the superior, so that the imagination and the will might be forced to believe the opposite of what the intellect understands. I am referring at all times to merely physical propositions, and not to supernatural things which are matters of faith.

I entreat those wise and prudent Fathers to consider with great care the difference that exists between doctrines subject to proof and those subject to opinion. Considering the force exerted by logical deductions, they may ascertain that it is not in the power of the professors of demonstrative sciences to change their opinions at will and apply themselves first to one side and then to the other. There is a great difference between commanding a mathematician or a philosopher and influencing a lawyer or a merchant, for demonstrated conclusions about things in nature or in the heavens cannot be changed with the same facility as opinions about what is or is not lawful in a contract, bargain, or bill of exchange. [...]

Now if truly demonstrated physical conclusions need not be subordinated to biblical passages, but the latter must rather be shown not to interfere with the former, then before a physical proposition is condemned it must be shown to be not rigorously demonstrated – and this is to be done not by those who hold the proposition to be true, but by those who judge it to be false. This seems very reasonable and natural, for those who believe an argument to be false may much more easily find the fallacies in it than men who consider it to be true and conclusive. Indeed, in the latter case it will happen that the more the adherents of an opinion turn over their pages, examine the arguments, repeat the observations, and compare the experiences, the more they will be confirmed in that belief. And Your Highness knows what happened to the late mathematician of the University of Pisa who undertook in his old age to look into the Copernican doctrine in the hope of shaking its foundations and refuting it, since he considered it false only because he had never studied it. As it fell out, no sooner had he understood its grounds, procedures, and demonstrations than he found himself persuaded, and from an opponent he became a very staunch defender of it. I might also name other mathematicians who, moved by my latest discoveries, have confessed it necessary to alter the previously accepted system of the world, as this is simply unable to subsist any longer.

If in order to banish the opinion in question from the world it were sufficient to stop the mouth of a single man – as perhaps those men persuade themselves who, measuring the minds of others by their own, think it impossible that this doctrine should be able to continue to find adherents – then that would be very easily done. But things stand otherwise. To carry out such a decision it would be necessary not only to prohibit the book of Copernicus and the writings of other authors who follow the same opinion, but to ban the whole science of astronomy. Furthermore, it would be necessary to forbid men to look at the heavens, in order that they might not see Mars and Venus sometimes quite near the earth and sometimes very distant, the variation being so great that Venus is forty times and Mars sixty times as large at one time as another. And it would be necessary to prevent Venus being seen round at one time and forked at another, with very thin horns; as well as many other sensory observations which can never be reconciled with the Ptolemaic system in any way, but are very strong arguments for the Copernican. And to ban Copernicus now that his doctrine is daily reinforced by many new observations and by the learned applying themselves to the reading of his book, after this opinion has been allowed and tolerated for those many years during which it was less followed and less confirmed, would seem in my judgment to be a contravention of truth, and an attempt to hide and suppress her the more as she revealed herself the more clearly and plainly. Not to abolish and censure his whole book, but only to condemn as erroneous this particular proposition, would (if I am not mistaken) be a still greater detriment to the minds of men, since it would afford them occasion to see a proposition proved that it was heresy to believe. [. . .]

Regarding the state of rest or motion of the sun and earth, experience plainly proves that in order to accommodate the common people it was necessary to assert of these things precisely what the words of the Bible convey. Even in our own age, people far less primitive continue to maintain the same opinion for reasons which will be found extremely trivial if well weighed and examined, and upon the basis of experiences that are wholly false or altogether beside the point. Nor is it worth while to try to change their opinion, they being unable to understand the arguments on the opposite side, for these depend upon observations too precise and demonstrations too subtle, grounded on abstractions which require too strong an imagination to be comprehended by them. Hence even if the stability of heaven and the motion of the earth should be more than certain in the minds of the wise, it would still be necessary to assert the contrary for the preservation of belief among the all-too-numerous vulgar. Among a thousand ordinary men who might be questioned concerning these things, probably not a single one will be found to answer anything except that it looks to him as if the sun moves and the earth stands still, and therefore he believes this to be certain. But one need not on that account take the common popular assent as an argument for the truth of what is stated; for if we should examine these very men concerning their reasons for what they believe, and on the other hand listen to the experiences and proofs which induce a few others to believe the contrary, we should find the latter to be persuaded by very sound arguments, and the former by simple appearances and vain or ridiculous impressions.

It is sufficiently obvious that to attribute motion to the sun and rest to the earth was therefore necessary lest the shallow minds of the common people should become confused, obstinate, and contumacious in yielding assent to the principal articles that are absolutely matters of faith. And if this was necessary, there is no wonder at all that it was carried out with great prudence in the holy Bible. I shall say further that not only respect for the incapacity of the vulgar, but also current opinion in those times, made the sacred authors accommodate themselves (in matters unnecessary to salvation) more to accepted usage than to the true essence of things. [...]

(b) *Cardinal Bellarmine, letter to Paolo Foscarini, 12 April 1615*

I have gladly read the letter in Italian and the essay in Latin that Your Reverence has sent me, and I thank you for both, confessing that they are filled with ingenuity and learning. But since you ask for my opinion, I shall give it to you briefly, as you have little time for reading and I for writing.

First. I say that it appears to me that Your Reverence and Sig. Galileo did prudently to content yourselves with speaking hypothetically and not positively, as I have always believed Copernicus did. For to say that assuming the earth moves and the sun stands still saves all the appearances better than eccentrics and epicycles is to speak well. This has no danger in it, and it suffices for mathematicians. But to wish to affirm that the sun is really fixed in the centre of the heavens and merely turns upon itself without travelling from east to west, and that the earth is situated in the third sphere and revolves very swiftly around the sun, is a very dangerous thing, not only by irritating all the theologians and scholastic philosophers, but also by injuring our holy faith and making the sacred Scripture false. For your Reverence has indeed demonstrated many ways of expounding the Bible, but you have not applied them specifically, and doubtless you would have had a great deal of difficulty if you had tried to explain all the passages that you yourself had cited.

Second. I say that, as you know, the Council of Trent would prohibit expounding the Bible contrary to the common agreement of the holy Fathers. And if Your Reverence would read not only all their works but the commentaries of modern writers on Genesis, Psalms, Ecclesiastes, and Joshua, you would find that all agree in expounding literally that the sun is in the heavens and travels swiftly around the earth, while the earth is far from the heavens and remains motionless in the centre of the world. Now consider whether, in all prudence, the Church could support the giving to Scripture of a sense contrary to the holy Fathers and all the Greek and Latin expositors. Nor may it be replied that this is not a matter of faith, since if it is not so with regard to the subject matter, it is with regard to those who have spoken. Thus that man would be just as much a heretic who denied that Abraham had two sons and Jacob twelve, as one who denied the virgin birth of Christ, for both are declared by the Holy Ghost through the mouths of the prophets and apostles.

Third. I say that if there were a true demonstration that the sun was in the centre of the universe and the earth in the third sphere, and that the sun did not go around the earth but the earth went around the sun, then it would be necessary to use careful consideration in explaining the Scriptures that seemed contrary, and we should rather have to say that we do not understand them than to say that something is false which had been proven. But I do not think there is any such demonstration, since none has been shown to me. To demonstrate that the appearances are saved by assuming the sun at the centre and the earth in the heavens is not the same thing as to demonstrate that in fact the sun is in the centre and the earth in the heavens. I believe that the first demonstration may exist, but I have very grave doubts about the second; and in case of doubt one may not abandon the Holy Scriptures as expounded by the holy Fathers. I add that the words 'The sun also riseth and the sun goeth down, and hasteth to the place where he ariseth' (Ecclesiastes 1:5) were written by Solomon, who not only spoke by divine inspiration, but was a man wise above all others of all created things, which wisdom he had from God; so it is not very likely that he would affirm something that was contrary to demonstrated truth, or truth that might be demonstrated. And if you tell me that Solomon spoke according to the appearances, and that it seems to us that the sun goes round when the earth turns, as it seems to one aboard ship that the beach moves away, I shall answer thus. Anyone who departs from the beach, though to him it appears that the beach moves away, yet knows that this is an error and corrects it, seeing clearly that the ship moves and not the beach; but as to the sun and earth, no sage has needed to correct the error, since he clearly experiences that the earth stands still and that his eye is not deceived when it judges the sun to move, just as he is likewise not deceived when it judges that the moon and the stars move. And that is enough for the present.

4.2 Galileo, *The Assayer*, 1623, in *Discoveries and Opinions of Galileo*, trans. Stillman Drake (New York: Doubleday, 1957), pp. 274–7

[...] But first I must consider what it is that we call heat, as I suspect that people in general have a concept of this which is very remote from the truth. For they believe that heat is a real phenomenon, or property, or quality, which actually resides in the material by which we feel ourselves warmed. Now I say that whenever I conceive any material or corporeal substance, I immediately feel the need to think of it as bounded, and as having this or that shape; as being large or small in relation to other things, and in some specific place at any given time; as being in motion or at rest; as touching or not touching some other body; and as being one in number, or few, or many. From these conditions I cannot separate such a substance by any stretch of my imagination. But that it must be white or red, bitter or sweet, noisy or silent, and of sweet or foul odor, my mind does not feel compelled to bring in as necessary accompaniments. Without the senses as our guides, reason or imagination unaided would probably never arrive at qualities like these. Hence I think that tastes, odors, colors, and so on are no more than mere names so far as the object in which we place them is concerned, and that they reside only in the consciousness. Hence if the living creature were removed, all these qualities would be wiped away and annihilated. But since we have imposed upon them special names, distinct from those of the other and real qualities mentioned previously, we wish to believe that they really exist as actually different from those.

I may be able to make my notion clearer by means of some examples. I move my hand first over a marble statue and then over a living man. As to the effect flowing from my hand, this is the same with regard to both objects and my hand; it consists of the primary phenomena of motion and touch, for which we have no further names. But the live body which receives these operations feels different sensations according to the various places touched. When touched upon the soles of the feet, for example, or under the knee or armpit, it feels in addition to the common sensation of touch a sensation on which we have imposed a special name, 'tickling'. This sensation belongs to us and not to the hand. Anyone would make a serious error if he said that the hand, in addition to the properties of moving and touching, possessed another faculty of 'tickling', as if tickling were a phenomenon that resided in the hand that tickled. A piece of paper or a feather drawn lightly over any part of our bodies performs intrinsically the same operations of moving and touching, but by touching the eye, the nose, or the upper lip it excites in us an almost intolerable titillation, even though elsewhere it is scarcely felt. This titillation belongs entirely to us and not to the feather; if the live and sensitive body were removed it would remain no more than a mere word. I believe that no more solid an existence belongs to many qualities which we have come to attribute to physical bodies – tastes, odors, colors, and many more.

A body which is solid and, so to speak, quite material, when moved in contact with any part of my person produces in me the sensation we call touch. This, though it exists over my entire body, seems to reside principally in the palms of the hands and in the finger tips, by whose means we sense the most minute differences in texture that are not easily distinguished by other parts of our bodies. Some of these sensations are more pleasant to us than others.... The sense of touch is more material than the other sense; and, as it arises from the solidity of matter, it seems to be related to the earthly element.

Perhaps the origin of two other senses lies in the fact that there are bodies which constantly dissolve into minute particles, some of which are heavier than air and descend, while others are lighter and rise up. The former may strike upon a certain part of our bodies that is much more sensitive than the skin, which does not feel the invasion of such subtle matter. This is the upper surface of the tongue; here the tiny particles are received, and mixing with and penetrating its moisture, they give rise to tastes, which are sweet or unsavory according to the various shapes, numbers, and speeds of the particles. And those minute particles which rise up may enter by our nostrils and strike upon some small protuberances which are the instrument of smelling; here likewise their touch and passage is received to our like or dislike according as they have this or that shape, are fast or slow, and are numerous or few. The tongue and nasal passages are providently arranged for these things, as the one extends from below to receive descending particles, and the other is adapted to those which ascend. Perhaps the excitation of tastes may be given a certain analogy to fluids, which descend through air, and odors to fires, which ascend.

Then there remains the air itself, an element available for sounds, which come to us indifferently from below, above, and all sides – for we reside in the air and its movements displace it equally in all directions. The location of the ear is most fittingly accommodated to all positions in space. Sounds are made and heard by us when the air – without any special property of 'sonority' or 'transonority' – is ruffled by a rapid tremor into very minute waves and moves certain cartilages of a tympanum in our ear. External means capable of thus ruffling the air are very numerous, but for the most part they may be reduced to the trembling of some body which pushes the air and disturbs it. Waves are propagated very rapidly in this way, and high tones are produced by frequent waves and low tones by sparse ones.

To excite in us tastes, odors, and sounds I believe that nothing is required in external bodies except shapes, numbers, and slow or rapid movements. I think that if ears, tongues, and noses were removed, shapes and numbers and motions would remain, but not odors or tastes or sounds. The latter, I believe, are nothing more than names when separated from living beings, just as tickling and titillation are nothing but names in the absence of such things as noses and armpits. . . .

Having shown that many sensations which are supposed to be qualities residing in external objects have no real existence save in us, and outside ourselves are mere names, I now say that I am inclined to believe heat to be of this character. Those materials which produce heat in us and make us feel warmth, which are known by the general name of 'fire', would then be a multitude of minute particles having certain shapes and moving with certain velocities. Meeting with our bodies, they penetrate by means of their extreme subtlety, and their touch as felt by us when they pass through our substance is the sensation we call 'heat'. This is pleasant or unpleasant according to the greater or smaller speed of these particles as they go pricking and penetrating; pleasant when this assists our necessary transpiration, and obnoxious when it causes too great a separation and dissolution of our substance. The operation of fire by means of its particles is merely that in moving it penetrates all bodies, causing their speedy or slow dissolution in proportion to the number and velocity of the fire-corpuscles and the density or tenuity of the bodies. . . .

4.3 MS G3 in the Archive of the Sacred Congregation for the Doctrine of the Faith, ser. AD EE [1624?], trans. P. Rosenthal in P. Redondi, *Galileo Heretic* (London: Allen Lane, 1988), pp. 333–5

Having in past days perused Signor Galileo Galilei's book entitled *The Assayer*, I have come to consider a doctrine already taught by certain ancient philosophers and effectively rejected by Aristotle, but renewed by the same Signor Galilei. And having decided to compare it with the true and undoubted Rule of revealed doctrines, I have found that in the Light of that Lantern which by the exercise and merit of our faith shines out indeed in murky places, and which more securely and more certainly than any natural evidence illuminates us, this doctrine appears false, or even (which I do not judge) very difficult and dangerous. So that he who receives the Rule as true must not falter in speech and in the judgment of more serious matters, I have therefore thought to propose it to you, Very Reverend Father, and beg you, as I am doing, to tell me its meaning, which will serve as my warning.

Therefore, the aforesaid Author, in the book cited (on page 196, line 29), wishing to explain that proposition proffered by Aristotle in so many places – that motion is the cause of heat – and to adjust it to his intention, sets out to prove that these accidents which are commonly called colors, odors, tastes, etc., on the part of the subject, in which it is commonly believed that they are found, are nothing but pure words and are only in the sensitive body of the animal that feels them. He explains this with the example of the

Tickle, or let us say Titillation, caused by touching a body in certain parts, concluding that like the tickle, as far as the action goes, once having removed the animal's sensitivity, it is no different from the touch and movement that one makes on a marble statue, for everything is our subjective experience; thus, these accidents which are apprehended by our senses and are called tastes, smells, colors, etc., are not, he says, subjects as one holds them generally to be, but only our senses, since the titillation is not in the hand or in the feather, which touches, for example, the sole of the foot, but solely in the animal's sensitive organ.

But this discourse seems to me to be at fault in taking as proved that which it must prove, i.e. that in all cases the object which we feel is in us, because the act that is involved is in us. It is the same as saying: the sight with which I see the light of the sun is in me; therefore, the light of the sun is in me. What might be the meaning of such reasoning, however, I shall not pause to examine.

The author then goes on to explain his Doctrine, and does his best to demonstrate what these accidents are in relation to the object and the end of our actions; and as one can see on page 198, line 12, he begins to explain them with the atoms of Anaxagoras or of Democritus, which he calls minims or minimal particles; and in these, he says continually, are resolved the bodies, which, however, applied to our senses penetrate our substance, and according to the diversity of the touches, and the diverse shapes of those minims, smooth or rough, hard or yielding, and according to whether they are few or many, prick us differently, and piercing with greater or lesser division, or by making it easier for us to breathe, and hence our irritation or pleasure. To the more material or corporeal sense of touch, he says, the minims of earth are most appropriate. To the taste, those of water and he calls them fluids; to the smell, those of fire and he calls them fiery particles; to the hearing, those of the air; and to the sight he then attributes the light, about which he says he has little to say. And on page 199, line 25, he concludes that in order to arouse in us tastes, smells, etc., all that is needed in bodies which commonly are tasteful, odorous, etc. are sizes, many varied shapes; and that the smells, tastes, colors, etc. are nowhere but in the eyes, tongues, noses, etc., so that once having taken away those organs, the aforesaid accidents are not distinguished from atoms except in name.

Now if one admits this philosophy of accidents as true, it seems to me, that makes greatly difficult the existence of the accidents of the bread and wine which in the Most Holy Sacrament are separated from their substance; since finding again therein the terms, and the objects of touch, sight, taste, etc., one will also have to say according to this doctrine that there are the very tiny particles with which the substance of the bread first moved our senses, which if they were substantial (as Anaxagoras said, and this author seems to allow on page 200, line 28), it follows that in the Sacrament there are substantial parts of bread or wine, which is the error condemned by the Sacred Tridentine Council, Session 13, Canon 2.

Or actually, if they were only sizes, shapes, numbers, etc., as he also seems clearly to admit, agreeing with Democritus, it follows that all these are accidental modes, or, as others say, shapes of quantity. While the Sacred Councils, and especially the Trident Council in the passage cited, determine that after the Consecration there remain in the Sacrament only the Accidents of the bread and wine, he instead says that there only remains the quantity with triangular shapes, acute or obtuse, etc., and that with these accidents alone is saved the existence of accidents or sensible species – which consequence seems to me not only in conflict with the entire communion of Theologians who teach us that in the Sacrament remain all the sensible accidents of bread, wine, color, smell, and taste, and not mere words, but also, as is known, with the good *judgment* that the quantity of the substance does not remain. Again, this is inevitably repugnant to the truth of the Sacred Councils; for, whether these minims are explained with Anaxagoras or Democritus, if they remain after the Consecration there will not be less substance of the bread in a consecrated host than in an unconsecrated host, since to be corporeal substance, in their opinion, consists, in an aggregation of atoms in this or that fashion, with this or that shape, etc. But if these particles do not remain, it follows that no accident of bread remains in the consecrated Host; since other accidents do not emerge, this Author says on page 197, line 1, that shapes, sizes, movements, etc. do so, and (these being the effects of a quantity or quantum substance) it is not possible, as all philosophers and Theologians teach, to separate them in such a way that they would exist without the substance or quantity of which they are accidents.

And this is what seems to me difficult in this Doctrine; and I propose and submit it, as regards my already expressed judgment, to what you, Most Reverend Father, will be pleased to tell and to which I make obeisance.

4.4 Galileo, *Dialogue Concerning the Two Chief World Systems, Ptolemaic and Copernican*, 1632, trans. Stillman Drake (Berkeley, CA: University of California Press, 1962), pp. 139, 141–2, 144–9

SALV. . . . Aristotle says, then, that a most certain proof of the earth's being motionless is that things projected perpendicularly upward are seen to return by the same line to the same place from which they were thrown, even though the movement is extremely high. This, he argues, could not happen if the earth moved, since in the time during which the projectile is moving upward and then downward it is separated from the earth, and the place from which the projectile began its motion would go a long way toward the east, thanks to the revolving of the earth, and the falling projectile would strike the earth that distance away from the place in question. [...]

SIMP.... Besides which, there is the very appropriate experiment of the stone dropped from the top of the mast of a ship, which falls to the foot of the mast when the ship is standing still, but falls as far from that same point when the ship is sailing as the ship is perceived to have advanced during the time of the fall, this being several yards when the ship's course is rapid.

SALV. There is a considerable difference between the matter of the ship and that of the earth under the assumption that the diurnal motion belongs to the terrestrial globe. For it is quite obvious that just as the motion of the ship is not its natural one, so the motion of all the things in it is accidental; hence it is no wonder that this stone which was held at the top of the mast falls down when it is set free, without any compulsion to follow the motion of the ship. But the diurnal rotation is being taken as the terrestrial globe's own and natural motion, and hence that of all its parts, as a thing indelibly impressed upon them by nature. Therefore the rock at the top of the tower has as its primary tendency a revolution about the center of the whole in twenty-four hours, and it eternally exercises this natural propensity no matter where it is placed. [...]

Now tell me: If the stone dropped from the top of the mast when the ship was sailing rapidly fell in exactly the same place on the ship to which it fell when the ship was standing still, what use could you make of this falling with regard to determining whether the vessel stood still or moved?

SIMP. Absolutely none....

SALV. Very good. Now, have you ever made this experiment of the ship?

SIMP. I have never made it, but I certainly believe that the authorities who adduced it had carefully observed it. Besides, the cause of the difference is so exactly known that there is no room for doubt.

SALV. You yourself are sufficient evidence that those authorities may have offered it without having performed it, for you take it as certain without having done it, and commit yourself to the good faith of their dictum. Similarly it not only may be, but must be that they did the same thing too – I mean, put faith in their predecessors, right on back without ever arriving at anyone who had performed it. For anyone who does will find that the experiment shows exactly the opposite of what is written; that is, it will show that the stone always falls in the same place on the ship, whether the ship is standing still or moving with any speed you please. Therefore, the same cause holding good on the earth as on the ship, nothing can be inferred about the earth's motion or rest from the stone falling always perpendicularly to the foot of the tower.

SALV. If you had referred me to any other agency than experiment, I think that our dispute would not soon come to an end; for this appears to me to be a thing so remote from human reason that there is no place in it for credulity or probability.

SALV. For me there is, just the same.
SIMP. So you have not made a hundred tests, or even one? And yet you so freely declare it to be certain? I shall retain my incredulity, and my own confidence that the experiment has been made by the most important authors who make use of it, and that it shows what they say it does.
SALV. Without experiment, I am sure that the effect will happen as I tell you, because it must happen that way; and I might add that you yourself also know that it cannot happen otherwise, no matter how you may pretend not to know it – or give that impression. But I am so handy at picking people's brains that I shall make you confess this in spite of yourself.

Sagredo is very quiet; it seemed to me that I saw him move as though he were about to say something.
SAGR. I was about to say something or other, but the interest aroused in me by hearing you threaten Simplicio with this sort of violence in order to reveal the knowledge he is trying to hide has deprived me of any other desire; I beg you to make good your boast. [...]
SALV. I do not want you to declare or reply anything that you do not know for certain. Now tell me: Suppose you have a plane surface as smooth as a mirror and made of some hard material like steel. This is not parallel to the horizon, but somewhat inclined, and upon it you have placed a ball which is perfectly spherical and of some hard and heavy material like bronze. What do you believe this will do when released? Do you not think, as I do, that it will remain still?
SIMP. If that surface is tilted?
SALV. Yes, that is what was assumed.
SIMP. I do not believe that it would stay still at all; rather, I am sure that it would spontaneously roll down.
SALV. Pay careful attention to what you are saying, Simplicio, for I am certain that it would stay wherever you placed it.
SIMP. Well, Salviati, so long as you make use of assumptions of this sort I shall cease to be surprised that you deduce such false conclusions.
SALV. Then you are quite sure that it would spontaneously move downward?
SIMP. What doubt is there about this? [...]
SALV.... Now how long would the ball continue to roll, and how fast? Remember that I said a perfectly round ball and a highly polished surface, in order to remove all external and accidental impediments. Similarly I want you to take away any impediment of the air caused by its resistance to separation, and all other accidental obstacles, if there are any.
SIMP. I completely understood you, and to your question I reply that the ball would continue to move indefinitely, as far as the slope of the surface extended, and with a continually accelerated motion....
SALV. But if one wanted the ball to move upward on this same surface, do you think it would go?

SIMP. Not spontaneously, no; but drawn or thrown forcibly, it would.
SALV. And if it were thrust along with some impetus impressed forcibly upon it, what would its motion be, and how great?
SIMP. The motion would constantly slow down and be retarded, being contrary to nature, and would be of longer or shorter duration according to the greater or lesser impulse and the lesser or greater slope upward.
SALV. Very well....

Now tell me what would happen to the same movable body placed upon a surface with no slope upward or downward.
SIMP. Here I must think a moment about my reply. There being no downward slope, there can be no natural tendency toward motion; and there being no upward slope, there can be no resistance to being moved, so there would be an indifference between the propensity and the resistance to motion. Therefore it seems to me that it ought naturally to remain stable....
SALV. I believe it would do so if one set the ball down firmly. But what would happen if it were given an impetus in any direction?
SIMP. It must follow that it would move in that direction.
SALV. But with what sort of movement? One continually accelerated, as on the downward plane, or increasingly retarded as on the upward one?
SIMP. I cannot see any cause for acceleration or deceleration, there being no slope upward or downward.
SALV. Exactly so. But if there is no cause for the ball's retardation, there ought to be still less for its coming to rest; so how far would you have the ball continue to move?
SIMP. As far as the extension of the surface continued without rising or falling.
SALV. Then if such a space were unbounded, the motion on it would likewise be boundless? That is, perpetual?
SIMP. It seems so to me, if the movable body were of durable material.
SALV. That is of course assumed, since we said that all external and accidental impediments were to be removed, and any fragility on the part of the moving body would in this case be one of the accidental impediments.

Now tell me, what do you consider to be the cause of the ball moving spontaneously on the downward inclined plane, but only by force on the one tilted upward?
SIMP. That the tendency of heavy bodies is to move toward the center of the earth, and to move upward from its circumference only with force; now the downward surface is that which gets closer to the center, while the upward one gets farther away.
SALV. Then in order for a surface to be neither downward nor upward, all its parts must be equally distant from the center. Are there any such surfaces in the world?
SIMP. Plenty of them; such would be the surface of our terrestrial globe if it were smooth, and not rough and mountainous as it is. But there is that of the water, when it is placid and tranquil.

SALV. Then a ship, when it moves over a calm sea, is one of these movables which courses over a surface that is tilted neither up nor down, and if all external and accidental obstacles were removed, it would thus be disposed to move incessantly and uniformly from an impulse once received?
SIMP. It seems that it ought to be.
SALV. Now as to that stone which is on top of the mast; does it not move, carried by the ship, both of them going along the circumference of a circle about its center? And consequently is there not in it an ineradicable motion, all external impediments being removed? And is not this motion as fast as that of the ship?
SIMP. All this is true, but what next?
SALV. Go on and draw the final consequence by yourself, if by yourself you have known all the premises.
SIMP. By the final conclusion you mean that the stone, moving with an indelibly impressed motion, is not going to leave the ship, but will follow it, and finally will fall at the same place where it fell when the ship remained motionless. And I, too, say that this would follow if there were no external impediments to disturb the motion of the stone after it was set free. But there are two such impediments; one is the inability of the movable body to split the air with its own impetus alone, once it has lost the force from the oars which it shared as part of the ship while it was on the mast; the other is the new motion of falling downward, which must impede its other, forward, motion.
SALV. As for the impediment of the air, I do not deny that to you, and if the falling body were of very light material, like a feather or a tuft of wool, the retardation would be quite considerable. But in a heavy stone it is insignificant, and if, as you yourself just said a little while ago, the force of the wildest wind is not enough to move a large stone from its place, just imagine how much the quiet air could accomplish upon meeting a rock which moved no faster than the ship! All the same, as I said, I concede to you the small effect which may depend upon such an impediment, just as I know you will concede to me that if the air were moving at the same speed as the ship and the rock, this impediment would be absolutely nil.

As for the other, the supervening motion downward, in the first place it is obvious that these two motions (I mean the circular around the center and the straight motion toward the center) are not contraries, nor are they destructive of one another, nor incompatible. As to the moving body, it has no resistance whatever to such a motion, for you yourself have already granted the resistance to be against motion which increases the distance from the center, and the tendency to be toward motion which approaches the center. From this it follows necessarily that the moving body has neither a resistance nor a propensity to motion which does not approach toward or depart from the center, and in consequence no cause for diminution in the property impressed upon it. Hence the cause of motion is not a single one which must be weakened by the new action, but there exist two distinct causes. Of these, heaviness attends only to the drawing of the movable body toward the center, and impressed force only to its being led around the center, so no occasion remains for any impediment. [...]

4.5 Tommasso Campanella, *Civitas Solis* (*City of the Sun*), 1623, trans. and intro. Daniel Donno (Berkeley, CA: University of California Press, 1981), pp. 27–37, 43–5

THE CITY OF THE SUN

A POETICAL DIALOGUE

INTERLOCUTORS

A Knight Hospitaler and a Genoese, one of Columbus' sailors

HOSPITALER. Tell me, please, all that happened to you on this voyage.

GENOESE. I have already told you how I sailed around the world and came to Taprobana, where I was forced to put ashore, how I hid in a forest to escape the fury of the natives, and how I came out onto a great plain just below the equator.

HOSPITALER. What happened to you there?

GENOESE. I soon came upon a large company of armed men and women, and many of them understood my language. They led me to the City of the Sun.

HOSPITALER. Tell me, what is that city like, and how is it ruled?

GENOESE. Rising from a broad plain, there is a hill upon which the greater part of the city is situated, but its circling walls extend far beyond its base, so that the entire city is two miles and more in diameter and has a circumference of seven miles; but because it is on a rise, it contains more habitations than it would if it were on a plain.

The city is divided into seven large circuits, named after the seven planets. Passage from one to the other is provided by four avenues and four gates facing the four points of the compass. [...]

At the summit of the hill there is a spacious plain in the center of which rises an enormous temple of astonishing design.

HOSPITALER. Tell me more, I beg you, tell me more.

GENOESE. The temple is perfectly circular and has no enclosing walls. It rests on large, well-proportioned columns. The large dome has a cupola at its center with an aperture directly above the single altar in the middle of the temple. The columns are arranged in a circle having a circumference of three hundred paces or more. Eight paces beyond them are cloisters with walls scarcely rising above the benches which are arranged along the concave exterior wall. Among the interior columns, which support the temple with no interposing walls, there are a large number of portable chairs.

Nothing rests on the altar but a huge celestial globe, upon which all the heavens are described, with a terrestrial globe beside it. On the vault of the dome overhead appear all the larger stars with their names and the influences they each have upon earthly things set down in three verses. The poles and circles are indicated, but not entirely since there is no wall below. Instead they are completed on the globes resting on the altar below. Seven lamps, each named for one of the seven planets, are always kept burning....

HOSPITALER. In good faith, tell me the manner of government you found among these people.

GENOESE. They have a Prince Prelate among them whom they call Sun, but in our language he would be called Metaphysician. He is both their spiritual and their temporal chief, and all decisions terminate with him.

There are also three collateral princes: Pon, Sin, and Mor, that is to say Power, Wisdom, and Love.

Power has charge of war and peace and of military affairs. He is supreme in war, but not above Sun. He has charge over officers, warriors, soldiers, munitions, fortifications, and sieges.

Wisdom has charge of all the sciences and of all the doctors and masters of the liberal and mechanical arts. Below him there are as many officers as there are sciences. There is an Astrologer, a Cosmographer, a Geometer, a Logician, a Rhetorician, a Grammarian, a Physician, a Physical Scientist, a Politician, and a Moralist. Wisdom has but one book in which all the sciences are treated and which is taught to all the people after the manner of the Pythagoreans. He has had all of the sciences pictured on all of the walls and on the ravelins, both inside and out.

On the exterior walls of the temple, on the curtains which are let down when there is preaching so that it may be heard, all the stars are drawn in order, with three descriptive verses assigned to each one.

On the inner wall of the first circuit, all the mathematical figures – more than Euclid or Archimedes speaks of – are shown in their significant propositions. On the outer wall there is a map of the entire world with charts for each country setting forth their rites, customs, and laws; and the alphabet of each is inscribed above the native one.

On the inner wall of the second circuit there are both samples and pictures of all minerals, metals, and stones, both precious and nonprecious, with two descriptive verses for each one. On the outer wall all kinds of lakes, seas, rivers, wines, oils, and other liquids are shown with their sources of origin, their powers, and their qualities indicated. There are also carafes full of diverse liquids, a hundred and even three hundred years old, with which nearly all infirmities are cured.

On the inner wall of the third circuit every kind of herb and tree to be found in the world is represented. Moreover, specimens of each are grown in earthen vessels placed on the ravelins with explanations as to where they were first discovered, what their specific powers are, what their relation is to the stars, to metals, to parts of the body, and how they are used in medicine. On the outer wall are shown all manner of fish to be found in river, lake or ocean; their particular qualities; the way they live, breed, develop; their use; their correspondence to celestial and earthly things, to the arts, and to nature. I was astonished when I saw bishop fish, chain fish, nail fish, and starfish exactly resembling such things among us. There are sea urchins and molluscs, and all that is worth knowing about them is marvelously set down in word and picture.

On the inner wall of the fourth circuit are depicted all kinds of birds, their characteristics, sizes, and habits, and the Phoenix appears most real among them. On the outer wall are found all sorts of reptiles, serpents, dragons, worms, insects, flies, gnats, etc., with their habits, venoms, and attributes explained. These are more numerous than anyone thinks.

On the inner wall of the fifth circuit appear the perfect animals of the earth in such great variety as to amaze you. We do not know the thousandth part of them. Because these are large in body, they appear on the outer ravelins as well. How many horses alone there are of different kinds, all of them beautifully and accurately represented!

On the inner wall of the sixth circuit all the mechanical arts are displayed together with their inventors, their diverse forms, and their diverse uses in different parts of the world. On the outer wall all the founders of laws and of sciences and inventors of weapons appear. I found Moses, Osiris, Jupiter, Mercury, Muhammad, and many others there. In a place of special honor I saw Jesus Christ and the twelve Apostles, whom they hold in great regard. I saw Caesar, Alexander, Pyrrhus, and all the Romans. At this, when I marveled that they knew the histories of these men, they explained to me that they understood the languages of all of the nations and that they dispatched ambassadors throughout the world to learn what was both good and bad in each of them. They profit a good deal by doing this. I noted that explosives and printing were known in China before they became known among us. They have teachers for these things, and, without effort, merely while playing, their children come to know all the sciences pictorially before they are ten years old.

Love has charge of breeding and sees to the coupling of males and females who will produce healthy offspring. They laugh at us because we are careful about the breeding of dogs and horses while we pay no attention to our own breeding. He has charge of education, of medicines, of drugs, of the sowing and harvesting of crops, of the commissaries, and, in short, of all things pertaining to food, dress, and sexual intercourse. He has many men and women serving under him who are skilled in these functions.

The Metaphysician governs all matters through these three officers. Without him nothing is done. Thus, everything is discussed among the four together, and what the Metaphysician decides all agree to. [...]

The officials are chosen by the four leaders [i.e., Sun, Pon, Sin and Mor] and by the teachers of the various arts. These know very well who is most suited for the particular task or virtue over which he is to preside. The candidates are nominated in council, and everyone present may tell what he knows against them. However, no one can be elected Sun unless he knows the history of all the peoples – their ceremonies, rites, and governments – and the inventors of all the arts and laws. He must moreover know all the mechanical arts, but each of these can be learned in two days, thanks to the fact that they are practiced and are graphically described on the walls besides. In addition, he must know the mathematical, physical, and astrological sciences. But he need not be concerned with languages since they have interpreters who serve as their grammarians. Above all, he must be a metaphysician and theologian who understands the theory and practice of every art and every science, the similitudes and differences among things, the Necessity, Fate, and Harmony of the world, the Power, Wisdom, and Love of God and of all things, the degrees of being and their correspondence to celestial, terrestrial, and marine things; and he must study astrology and the prophets carefully. They know whom they are to elect Sun, therefore, but no one can have the post unless he is thirty-five years old. Once appointed, his tenure lasts until someone with greater knowledge and greater ability to rule is discovered. [...]

Chapter Five
Iberian Science: Navigation, Empire and Counter-Reformation

5.1 Garcia d'Orta, *Colloquies on the Simples and Drugs of India*, 1563, trans. C. Markham (London: Henry Sotheran and Co., 1913), pp. 368–73

RUANO

Tell me the appearance of the tree, how it grows and how all is grown on one tree, for in this Greeks, Latins, and Arabs all agree, as well as the writers who have treated of the subject recently.

ORTA

All agree, with one accord, not to tell the truth, although Dioscorides may be pardoned because he wrote with false information and at a great distance, with intervening seas not navigated as they now are. He was copied by Pliny, Galen, Isidore, Avicenna, and all the Arabs. But those who write now, such as Antonio Musa and the Friars, have the greater fault, because they merely repeat in the same way without taking the trouble to ascertain things so well known as the appearance of the tree, pepper, the fruit, how it ripens and how it is gathered.

RUANO

Are all those you have mentioned in error?

ORTA

Yes, if you call saying what is not true an error.

RUANO

This being so, tell me what you have seen or heard from persons worthy of belief, and afterwards I will come with my doubts.

ORTA

The tree of the pepper is planted at the foot of another tree, generally at the foot of a palm or cachou tree. It has a small root, and grows as its supporting tree grows, climbing round and embracing it. The leaves are not numerous, nor large, smaller than an orange leaf, green, and sharp pointed, burning a little almost like betel. It grows in bunches like grapes, and only differs in the pepper being smaller in the grains, and the bunches being smaller, and always green at the time that the pepper dries. The crop is in its perfection in the middle of January. In Malabar the plant is of two kinds, one being the black pepper and the other white; and besides these there is another in Bengal called the long pepper.

RUANO

It seems to me that you abolish all the writers, ancient and modern, by this that I have heard you say. For Dioscorides says that the tree of the pepper is low, and produces a long fruit like a sheath, which they call long pepper, and inside this sheath there are small grains like gram, and that this is the perfect pepper, for at the proper time these sheaths open and discover some close clusters and the grains which we know, and that they are gathered before they are quite ripe. He says that they are sour and these are the white pepper. They are ingredients of the medicines they make for sore eyes and against poison that has been drunk, and against the bites of venomous beasts. The long pepper is strongly biting and rather bitter, owing to having been gathered before it is ripe, and is therefore efficacious for the things I mentioned. The black pepper is more suave and sharper, and more agreeable to the taste, from having been gathered at the right time, and also more aromatic than the white kind, and so it is more profitable for tempering the food. The weakest of all is the white pepper, owing to having been gathered before it is ripe. The black pepper is heavier and better. The people of the country call it BARCAMANSI because some empty grains are found amongst it. This is what Dioscorides says on the subject, at present it being unnecessary to enter upon medicinal qualities. At the end of chapter x. he says that the root is like that of *costo*. Pliny says that the trees are like junipers, and that they grow only on Mount Caucasus according to what some say, also that the seeds are like those of the juniper, and that one seed divides or goes apart from another in a small part of the pod, like figs. The prices of them was 25 *livras* for long pepper, of black 16 to 18, white 17, a *livra* equal to 3 *cruzados*. He says that pepper in its own country is wild and not planted, and that in Italy he heard of a tree which was like a myrtle, also that there is pepper in the part of Arabia called Trogoldita, which is called in the language of that country BARCAMANSI. Everything else about its use is copied from Dioscorides, so it need not be referred to here. Avicenna has two chapters, one on FULFUL, the other on DARFULFUL, which is the long pepper, and both Avicenna and Galen do no more than copy from Dioscorides, and so with Serapiam, who only has what he found in Dioscorides and Galen. Something that Paulo Egineta wrote is not relevant. These are the remarks on the subject made by the ancients. Turning to St. Isidore, he must, as a saint, be considered a high authority. He says that when the people of the country find that the pepper is ripe for gathering, they set fire to the wood for fear of serpents and burn the serpents. The pepper turns black owing to the fire applied to the wood. But I, to tell you the real truth, look upon this as a fable; so I wrote it first and then spoke. St. Isidore cannot have said this because he believed it, but to relate what others said. So I do not care to make excuses about these things, for I do not believe them. But I must tell you that I do not know for what reason you discredit such ancient doctors, and of such high authority, whose statements are confirmed by modern writers such as Mateas Silvatico, Sepúlveda, Antonio Musa, the Spanish Friar, the Italian Friar, and so many others who have written on pharmacy. On this account I require you, in the name of God, to tell me only what you have seen and heard from persons well worthy of credence, confirming what you say by reasons which you know so well how to give, and finally we will consider how it is used in medicine by the physicians of this land, then I will put any necessary questions to you, and I regret if I have spoken too freely.

ORTA

In the first place, your worship must understand that pepper does not grow either on the skirt or on the slope of Mount Caucasus, as Pliny says. For there the price of pepper is higher than in any other country. This you must know, for you know how far Mount Caucasus is from Malabar or Sumatra, places where there is the greatest quantity of pepper. Nor is it like the juniper, for it is a climbing plant, while the juniper stands by itself, nor are its leaves like those of the juniper. Their shape is as I have already described to you, and the bunches grow like those of grapes. When they are green, with the berries apart and unripe, they put them into vinegar and salt. This I know very well from the testimony of my eyes. In the same way I know that the tree of long pepper grows in a land very distant from Malabar, the nearest point being 500 leagues off, for it is in Bengal and in Java. This long pepper is worth at Cochin, where there is the greatest quantity of black pepper, 5 *cruzados* the *quintal*, and four years ago at that place, when there was a greater demand for long pepper at other places, the *quintal* was worth 15 to 20 *cruzados*. The usual price of black pepper at Cochin is $2\frac{1}{2}$ *cruzados*, but in Bengal 12 *cruzados*; while the long pepper sells in Bengal at $1\frac{1}{2}$ *cruzados*. This is enough to show that the long pepper does not come from the same tree as the ordinary kind, much less is it needful for a man who has seen a thing with his eyes to give further proof of it. The white pepper comes from a tree of its own, and, to tell you the truth, there are not many but very few in Malabar or in Malacca. They put this pepper on the tables of the lords as we put salt. It is esteemed in both parts of Malabar as good against poison and for the eyes. It would be well if all that Dioscorides said was as true as that this pepper is good against poison. You will now see that these three trees are different, namely those of long, black, and white pepper. The long pepper is called PIMPILIM in Bengal. The tree of the long is no more like that of the black pepper than a bean is like an egg. The black and white pepper trees are very like each other, and only the people of the country can tell them apart, just as we cannot tell the black from the white vines unless they are bearing grapes. If you do not want to believe me, believe in these three seeds, that one is of long, the other of black, the other of white pepper. As for pepper being called BARCAMANSI no such name has ever been heard of in any of these countries, nor anything like it.

RUANO

Truly I find myself corrected, as I do not see it as the others do, it being made so clear.

ORTA

You see here the green pepper grown in clusters on this branch of a tree, and you see there another done with vinegar and salt, which you should taste before all.

RUANO

I see it all well, and now that I am corrected I see that the new writers never investigate satisfactorily. Laguna complains of the Portuguese because they do not describe these things and only care about skinning and robbing the Indians.

5.2 Nicolás Monardes, *La Historia Medicinal de las cosas que se traen de nuestras Indias Occidentales que sirven en Medicina* (*Medicinal History of the Things Brought from our West Indies* that are of use in Medicine*), Seville, 1565–74, trans. D. C. Goodman

In the year 1492 we Spaniards with the guidance of Christopher Columbus, a native of Genoa, were led to the discovery of the West Indies, now known as the New World. Since then numerous other isles have been discovered and much of the continent, including New Spain and Peru. And these many provinces, kingdoms and cities have diverse customs and things have been found there that have never been seen in any other part of the world; and there are other things which exist here but in much greater abundance over there. This applies to gold, silver, pearls, emeralds, turquoise, and other precious stones, which now arrive from those parts in quantities far greater than what is found here; especially the gold and silver, whose value in millions is astounding, quite apart from the abundance of pearls supplied to the world. From those parts also are brought parrots, monkeys,

* 'West Indies' here does not signify just the territory we now understand by this term; it then meant the whole of Hispanic America, both mainland and islands.

griffins, lions, falcons, hawks, tigers, wool, cotton, cochineal, hides, sugar, copper, Brazil-wood, ebony and lapis lazuli. And such is the quantity of these things which arrive every year – almost one hundred ships laden with them – that it is an amazing and unbelievable abundance.

In addition to this wealth from our West Indies come many trees, plants, herbs, roots, juices, gums, fruits, seeds, liquors and stones which have great medicinal virtues. And in these have been discovered very great effects which are far more valuable than all of the forementioned items, because bodily health is more necessary and excellent than material possessions. ... And, as Aristotle says, it is not surprising that different plants and fruits are produced in different lands; or that trees, plants and fruits are found in one region which do not occur in another. The dictamnus grows only in Crete; mastic only in Cyprus; cinnamon, cloves, pepper and other spices only in the Moluccas. And there are many other things in various parts of the world which have never been known until our times, which the Ancients did not have, and which time, the discoverer of all things, has revealed to us to our great benefit.

And so discoveries by the Spanish of new regions, kingdoms and provinces have led to the supply of new medicines and remedies for the treatment of numerous diseases that would otherwise have remained incurable. Although these things are known to some, they are not known to all. Therefore I set out to write about all things coming from our West Indies which serve the art of medicine to remedy the illnesses and diseases which afflict us. This will be no small benefit to our contemporaries, and also to future generations. I will make the beginning for others to add to through increasing knowledge and experience.

In this city of Seville, which is the port and terminal for all of the West Indies, we are more familiar with these things than any other part of Spain, because everything arrives here first. I have benefited from this, and in the course of practising medicine in this city for 40 years, I have used the things brought from those regions and experimented with them on numerous patients, with diligence, circumspection and very great success.

5.3 Diego de Zúñiga, *In Job commentaria*, Toledo, 1584, pp. 205–7, trans. D. C. Goodman

'The earth moves from its place and its pillars tremble' [*Job* 9:6]. It seems that this difficult text can be clarified by the Pythagorean doctrine which asserts that the earth, by its own nature, moves and that in no other way can the very varied motions of the stars be explained. According to Plutarch this doctrine was accepted by Philolaus and Heraclides of Pontus ... And in our own time Copernicus has described the course of the planets in accordance with this doctrine. There is no doubt that his theory gives a much better and truer account of the positions of the planets than Ptolemy's *Almagest* or other theories. Ptolemy in fact could not explain the precession of the equinoxes; nor could he establish a definite beginning to the year, as he himself acknowledged in the second chapter of the third part of the *Almagest*... Also we now know that the sun is over 40,000 stades closer than the Ancients thought, so that neither Ptolemy nor other astronomers knew the cause of this precession. But Copernicus gives very convincing explanations of this and other phenomena by assuming that the earth is in motion. Nor does his theory conflict at all with what Solomon says in the book of *Ecclesiastes*: 'the earth abides forever'. The meaning of this is that while there is a succession on earth of epochs and generations of men, the earth remains unchanged. And indeed the text says: 'one generation passes away and another generation comes, but the earth abides forever'. The text [of *Job*] is not consistent with the immobile earth of the philosophers. Nor is it contradicted by other statements in this chapter of *Ecclesiastes* and in many others of the Holy Scriptures which refer to the movement of the sun, which Copernicus regards as the immobile centre of the universe.... There is no passage in the Holy Scriptures which speaks so clearly of an immobile earth than this one does of its mobility. [...]

5.4 Diego de Zúñiga, *Philosophia prima pars*, Toledo, 1597, pp. 229*v*–230*v*, trans. D. C. Goodman

As for the situation of the earth, this is a matter of greater difficulty, and there is no certainty about it however much the most learned men, Aristotle, Ptolemy and many other philosophers and astronomers have tried to demonstrate that the earth's orb is situated in the centre of the universe... because the magnitude of the heavens could be so great that the appearances viewed from earth would be the same whether the earth was situated at the centre or very distant from it. And this is clear from Copernicus' great treatise.... It can be supposed that the extent and elevation of the universe are greater than anyone has ever thought. Now we come to the state of the earth, on which there is great controversy amongst the learned; nevertheless the earth's state can be discussed with greater probability than its position. That the earth is not at rest but by its nature moves was asserted by Pythagoras and Philolaus.... In our own age the same was taught by Nicolaus Copernicus whose learned treatise based the arrangement of the universe on the multiple movement of the earth... But Aristotle, Ptolemy and other most expert philosophers and astronomers are of the opposite opinion and we follow them in this.... Some of the movements which Copernicus and others give to the earth are not problematic. But that is not the case with the alleged rotation of the entire earth in twenty-four hours, which seems to reduce the idea of the earth's motion to absurdity. Pythagoras was forced to adopt this view once he accepted that... the sun remained at rest at the centre; and since the sun does not move, the diurnal movement was attributed to the earth rotating about its centre. This was taught by Pythagoreans and Copernicus. Aristotle and

Ptolemy refute this motion with persuasive arguments. The circumference of the earth is 80,181 stades.... Consequently every point on the earth's surface would traverse so much space in a day and with such a rapid motion from west to east that it would exceed the speed of clouds, birds and all other things suspended in the air, and all of these would appear to move to the west. Similarly it would be much more difficult to throw a lance or a stone to the east than towards the west...and heavy objects thrown straight up would never return to the same place, if the earth moved with such impetus; yet if this is repeated a thousand times, the objects always fall perpendicularly down; therefore the earth does not move.

5.5 Gaspar de Quiroga, *Index et Catalogus Librorum prohibitorum*, Madrid, 1583, pp. *3v–4v*, trans. D. C. Goodman

...The following are all prohibited: books, treatises, documents, recipes and registers for invoking demons in any manner, whether by necromancy, hydromancy, pyromancy, aeromancy, onomancy, chiromancy and geomancy; or by writings on the magic art, witchcraft, omens, incantations, spells, circles, characters, seals, rings and figures.

Also prohibited are all books, treatises and writings which discuss or give rules, or expound the art or science of acquiring from the stars or from the lines on the hand knowledge of the future, which depends on man's free-will and chance.... It is forbidden for anyone to make predictions on such things. But this does not apply to those parts of astrology which concern general events of the world, nor those parts which teach us to know our inclinations, conditions and bodily qualities; nor to those parts of astrology which have a bearing on agriculture, navigation and medicine.... As for the conjurations and exorcism used against demons and storms, apart from what is authorised by Rome in prayer, only texts from ecclesiastical manuals may be used which have been inspected and approved by Ordinaries.

Chapter Six
Science from the Earth in Central Europe

6.1 Ulrich Rülein von Calw, 'On the Origin of Metals', in *Ein nützlic Bergbüchlein* [1500], trans. H. Clark Hoover and L. Henry Hoover, 1912, from Georgius Agricola, *De Re Metallica*, 1556 (New York: Dover reprint, 1950), pp. 44–6

The first chapter or first part; on the common origin of ore, whether silver, gold, tin, copper, iron, or lead ore, in which they all appear together, and are called by the common name of metallic ore. It must be noticed that for the washing or smelting of metallic ore, there must be the one who works and the thing that is worked upon, or the material upon which the work is expended. The general worker (efficient force) on the ore and on all things that are born, is the heavens, its movement, its light and influences, as the philosophers say. The influence of the heavens is multiplied by the movement of the firmaments and the movements of the seven planets. Therefore, every metallic ore receives a special influence from its own particular planet, due to the properties of the planet and of the ore, also due to properties of heat, cold, dampness, and dryness. Thus gold is of the Sun or its influence, silver of the Moon, tin of Jupiter, copper of Venus, iron of Mars, lead of Saturn, and quicksilver of Mercury. Therefore, metals are often called by these names by hermits and other philosophers. Thus gold is called the Sun, in Latin *Sol*, silver is called the Moon, in Latin *Luna*, as is clearly stated in the special chapters on each metal. Thus briefly have we spoken of the 'common worker' of metal and ore. But the thing worked upon, or the common material of all metals, according to the opinion of the learned, is sulphur and quicksilver, which through the movement and influence of the heavens must have become united and hardened into one metallic body or one ore. Certain others hold that through the movement and the influence of the heavens, vapours or *braden*, called mineral exhalations, are drawn up from the depths of the earth, from sulphur and quicksilver, and the rising fumes pass into the veins and stringers and are united through the effect of the planets and made into ore. Certain others hold that metal is not formed from quicksilver, because in many places metallic ore is found and no quicksilver. But instead of quicksilver they maintain a damp and cold and slimy material is set up on all sulphur which is drawn out from the earth, like your perspiration, and from that mixed with sulphur all metals are formed. Now each of these opinions is correct according to a good understanding and right interpretation; the ore or metal is formed from the fattiness of the earth as the material of the first degree (primary element), also the vapours or *braden* on the one part and the materials on the other part, both of which are called quicksilver. Likewise in the mingling or union of the quicksilver and the sulphur in the ore, the sulphur is counted the male and the quicksilver the female, as in the bearing or conception of a child. Also the sulphur is a special worker in ore or metal.

The second chapter or part deals with the general capacity of the mountain. Although the influence of the heavens and the fitness of the material are necessary to the formation of ore or metal, yet these are not enough thereto. But there must be adaptability of the natural vessel in which the ore is formed, such are the veins. [...] Also there must be a suitable place in the mountain which the veins and stringers can traverse.

6.2 Cornelius Agrippa, *De Occulta Philosophia* [1510], 1531, in P. Maxwell-Stuart, *The Occult in Early Modern Europe* (Basingstoke: Macmillan, 1999), pp. 96–7

It is clear that all inferior things are subject to higher and (as Proclus says) in a certain fashion each is present inside the other, i.e. the highest is in the lowest and the lowest in the highest. Thus, terrestrial things are in heaven, but in a causal and celestial way; and celestial things are on earth, but in a terrestrial way, that is to say, consistent with their intention. So we say that here on earth there exist certain things which pertain to the Sun and some which pertain to the Moon, because in them the Sun and Moon give rise to something of their own power. This is why things of this kind receive more workings and properties like those of the stars and signs under which they exist. Thus, we find out that things pertaining to the Sun have a relationship with the heart and head because of Leo (the house of the Sun) and Aries (in which the Sun is exalted). Things pertaining to Mars are ascribed to the head and testicles because of Aries and Scorpio; which is why people whose senses are staggering and who have a pain in the head because they are drunk with wine find immediate relief by plunging their testicles into cold water, or washing them thoroughly with vinegar.

But with regard to these inter-relationships, one must know how the human body is allotted to the planets and their signs. According to Arabic tradition, the Sun rules the brain, heart, thigh, marrow, right eye and vital spirit. Mercury governs the tongue, mouth, the other instruments or organs of the senses (internal as well as external), the hands, feet, legs, nerves and power of imagination. Saturn rules the spleen, stomach, bladder, womb, right ear and the power of making connections between things. Jupiter rules the liver and the fleshier part of the stomach, the belly and the navel, which is why ancient authors tell us that a replica of the navel was deposited in the Temple of Jupiter Ammon. Some writers also attribute to Jupiter the ribs, pubic bone, intestines, blood, arms, right hand, left ear and the power of the genitals. Others, however, set Mars in charge of the blood, veins, kidneys, gall-bag, nostrils, back, descent of the sperm and the power to be angry. Venus (some say) governs the kidneys, testicles, vulva, uterus, sperm and lust, along with the flesh, body-fat, stomach, pubic area, umbilical cord and everything such as that, serving the sexual act; and she is said to rule, in addition, the ossacrum, backbone, loins, head and the mouth which gives a kiss as a pledge of love. The Moon, even though she may lay claim to the whole body and its individual parts because of the variety of her signs, nevertheless has ascribed to her in particular the brain, lungs, marrow of the backbone, stomach, the menstrual fluids and all the waste matters of the body, the left eye and the power to grow.

Hermes says there are seven openings in the head of a living creature and that these are assigned to the seven planets: namely, the right ear to Saturn, the left to Jupiter, the right nostril to Mars, the left to Venus, the right eye to the Sun, the left to the Moon and the mouth to Mercury.

Individual signs of the zodiac look after their special parts of the body. So, Aries rules the head and face; Taurus the neck; Gemini the arms and shoulders; Cancer the chest, lungs, stomach and upper arms; Leo the heart, stomach, liver and back; Virgo the intestines and the bottom of the stomach; Libra the kidneys, thighs and buttocks; Scorpio the genitals, the vulva and the uterus; Sagittarius the thigh and groins; Capricorn the knees; Aquarius the legs and shins; and Pisces the feet....

Those things relating to Saturn cause sadness and depression; those relating to Jupiter are conducive to happiness and excellence; those relating to Mars to boldness, contention and anger; those relating to the Sun bestow glory, victory and wrath; those relating to Venus grant love, lust and ardent desire; those relating to Mercury grant eloquence; those relating to the Moon bring a conventional life. People's skills and characters are also allotted according to the planets. Saturn governs old men, monks, those given to depression, hidden treasures and those things which one acquires with difficulty and by means of long journeys. Jupiter has control over members of religious Orders, prelates, Kings, Dukes and material profit lawfully gained. Mars governs barbers, surgeons, doctors, executioners, butchers, provisioners, bakers, millers, soldiers and those who are everywhere called 'the sons of Mars'.

6.3 Georgius Agricola, 'On the Origin of Metals', in *De Ortu et Causis Subterraneorum*, 1546, trans. H. Clark Hoover and L. Henry Hoover, 1912, Book III of *De Re Metallica*, 1556 (New York: Dover reprint, 1950), p. 51

Having now refuted the opinions of others, I must explain what it really is from which metals are produced. The best proof that there is water in their materials is the fact that they flow when melted, whereas they are again solidified by the cold of air or water. This, however, must be understood in the sense that there is more water in them and less 'earth'; for it is not simply water that is their substance but water mixed with 'earth'. And such a proportion of 'earth' is in the mixture as may obscure the transparency of the water, but not remove the brilliance which is frequently in unpolished things. Again, the purer the mixture, the more precious the metal which is made from it, and the greater its resistance to fire. But what proportion of 'earth' is in each liquid from which a metal is made no mortal can ever ascertain, or still less explain, but the one God has known it, Who has given certain sure and fixed laws to nature for mixing and blending things together. It is a juice (*succus*) then, from which metals are formed; and this juice is created by various operations. Of these operations the first is a flow of water which softens the 'earth' or carries the 'earth' along with it, thus there is a mixture of 'earth' and water, then the power of heat works upon the mixtures so as to produce that kind of a juice. We have spoken of the substance of metals; we must now speak of their efficient cause. [...]

We do not deny the statement of Albertus Magnus that the mixture of 'earth' and water is baked by subterranean heat to a certain denseness, but it is our opinion that the juice so obtained is afterward solidified by cold so as to become a metal. [...] We grant, indeed, that heat is the efficient cause of a good mixture of elements, and also cooks this same mixture into a juice, but until this juice is solidified by cold it is not a metal. [...] This view of Aristotle is the true one. For metals melt through the heat and somehow become softened; but those which have become softened through heat are again solidified by the influence of cold, and, on the contrary, those which become softened by moisture are solidified by heat.

6.4 Georgius Agricola, 'On the Knowledge of the Miner', Book I of *De Re Metallica*, 1556, trans. H. Clark Hoover and L. Henry Hoover, 1912 (New York: Dover reprint, 1950), pp. 1–4

Many persons hold the opinion that the metal industries are fortuitous and that the occupation is one of sordid toil, and altogether a kind of business requiring not so much skill as labour. But as for myself, when I reflect carefully upon its special points one by one, it appears to be far otherwise. For a miner must have the greatest skill in his work, that he may know first of all what mountain or hill, what valley or plain, can be prospected most profitably, or what he should leave alone; moreover, he must understand the veins, stringers and seams in the rocks. Then he must be thoroughly familiar with the many and varied species of earths, juices, gems, stones, marbles, rocks, metals, and compounds. He must also have a complete knowledge of the method of making all underground works. Lastly, there are the various systems of assaying substances and of preparing them for smelting; and here again there are many altogether diverse methods. [...]

Furthermore, there are many arts and sciences of which a miner should not be ignorant. First there is Philosophy, that he may discern the origin, cause, and nature of subterranean things; for then he will be able to dig out the veins easily and advantageously, and to obtain more abundant results from his mining. Secondly, there is Medicine, that he may be able to look after his diggers and other workmen, that they do not meet with those diseases to which they are more liable than workmen in other occupations, or if they do meet with them, that he himself may be able to heal them or may see that the doctors do so. Thirdly follows Astronomy, that he may know the divisions of the heavens and from them judge the direction of the veins. Fourthly, there is the science of Surveying that he may be able to estimate how deep a shaft should be sunk to reach the tunnel which is being driven to it, and to determine the limits and boundaries in these workings, especially in depth. Fifthly, his knowledge of Arithmetical Science should be such that he may calculate the cost to be incurred in the machinery and the working of the mine. Sixthly, his learning must comprise Architecture, that he himself may construct the various machines and timber work required underground, or that he may be able to explain the method of the construction to others. Next, he must have knowledge of Drawing, that he can draw plans of his machinery. Lastly, there is the Law, especially that dealing with metals, that he may claim his own rights, that he may undertake the duty of giving others his opinion on legal matters, that he may not take another man's property and so make trouble for himself, and that he may fulfil his obligations to others according to the law. [...]

But let us now approach the subject we have undertaken. Since there has always been the greatest disagreement amongst men concerning metals and mining, some praising, others utterly condemning them, therefore I have decided that before imparting my instruction, I should carefully weigh the facts with a view to discovering the truth in this matter.

6.5 Georgius Agricola, 'On Assaying', Book VII of *De Re Metallica*, 1556, trans. H. Clark Hoover and L. Henry Hoover, 1912 (New York: Dover reprint, 1950), pp. 219–20, 222–4

BOOK VII.

Since the Sixth Book has described the iron tools, the vessels and the machines used in mines, this Book will describe the methods of assaying ores; because it is desirable to first test them in order that the material mined may be advantageously smelted, or that the dross may be purged away and the metal made pure. Although writers have mentioned such tests, yet none of them have set down the directions for performing them, wherefore it is no wonder that those who come later have written nothing on the subject. By tests of this kind miners can determine with certainty whether ores contain any metal in them or not; or if it has already been indicated that the ore contains one or more metals, the tests show whether it is much or little; the miners also ascertain by such tests the method by which the metal can be separated from that part of the ore devoid of it; and further, by these tests, they determine that part in which there is much metal from that part in which there is little. Unless these tests have been carefully applied before the metals are melted out, the ore cannot be smelted without great loss to the owners. [...] Metals, when they have been melted out, are usually assayed in order that we may ascertain what proportion of silver is in a *centumpondium* of copper or lead, or what quantity of gold is in one *libra* of silver; and, on the other hand, what proportion of copper or lead is contained in a *centumpondium* of silver, or what quantity of silver is contained in one *libra* of gold. And from this we can calculate whether it will be worth while to separate the precious metals from the base metals, or not. Further, a test of this kind shows whether coins are good or are debased; and readily detects silver, if the coiners have mixed more than is lawful with the gold; or copper, if the coiners have alloyed with the gold or silver more of it than is allowable. I will explain all these methods with the utmost care that I can.

The method of assaying ore used by mining people, differs from smelting only by the small amount of material used. [...]

Both processes, however, are carried out in the same way, for just as we assay ore in a little furnace, so do we smelt it in the large furnace. Also in both cases charcoal and not wood is burned. Moreover, in the crucible when metals are tested, be they gold, silver, copper, or lead, they are mixed in precisely the same way as they are mixed in the blast furnace when they are smelted. Further, those who assay ores with fire, either pour out the metal in a liquid state, or, when it has cooled, break the crucible and clean the metal from slag; and in the same way the smelter, as soon as the metal flows from the furnace into the forehearth, pours in cold water and takes the slag from the metal with a hooked bar. Finally, in the same way that gold and silver are separated from lead in a cupel, so also are they separated in the cupellation furnace.

It is necessary that the assayer who is testing ore or metals should be prepared and instructed in all things necessary in assaying, and that he should close the doors of the room in which the assay furnace stands, lest anyone coming at an inopportune moment might disturb his thoughts when they are intent on the work. It is also necessary for him to place his balances in a case, so that when he weighs the little buttons of metal the scales may not be agitated by a draught of air, for that is a hindrance to his work.

6.6 Paracelsus, 'The Physician's Remedies' [1520s–1530s], trans. Norbert Guterman, in Jolande Jacobi (ed.), *Paracelsus: Selected Writings* (London: Routledge and Kegan Paul, 1951), pp. 158–60*

What sense would it make or what would it benefit a physician if he discovered the origin of the diseases but could not cure or alleviate them? And since the fit manner of preparation is not to be found in pharmaceutics, we must explore further; that is to say, we must learn from alchemy. In it we find the true cause and everything that is needed. Although alchemy has now fallen into contempt and is even considered a thing of the past, the physician should not be influenced by such judgments. For many arts, such as astronomy, philosophy, and others, are also in disrepute. I am directing you, physicians, to alchemy for the preparation of the *magnalia*, for the production of the *mysteria*, for the preparation of the *arcana*, for the separation of the pure from the impure, to the end that you may obtain a flawless, pure remedy, God-given, perfect, and of certain efficacy, achieving the highest degree of virtue and power. For it is not God's design that the remedies should exist for us ready-made, boiled, and salted, but that we should boil them ourselves, and it pleases Him that we boil them and learn in the process, that we train ourselves in this art and are not idle on earth, but labour in daily toil. For it is we who must pray for our daily bread, and if He grants it to us, it is only through our labour, our skill and preparation.

The first and highest book of medicine is called *Sapientia*. Without this book no one will achieve anything fruitful.... For this book is God himself. In Him who has created all things lies also wisdom, and only He knows the primal cause of all things.... Although the remedy is given by nature... it must be revealed to us by the all-highest book, so that we may learn what is in it, how it is made, how it is obtained from the earth, and how and to what patients it should be administered....

The second book of medicine – of this too you must take note! – is the firmament. [...] Just as a man reads a book on paper, so the physician is compelled to spell out the stars of the firmament in order to know his conclusions. [...]

The book of medicine is nature itself. And just as you see yourself in a mirror, so you must rediscover all your sciences in nature, with exactly the same certainty and with as little illusion as when you see yourself in a mirror.

Marvellous virtues are inherent in the remedies. One would hardly believe that nature contained such virtues.... For only a great artist is able to discover them, not one who is only versed in books, but only one who has acquired his ability and skill through the experience of his hands.... It

* Note that the editor (Jacobi), in this extract and the following one, grouped together around themes a number of short passages taken from a wide variety of works composed at different times.

is an important art, and therefore it cannot be clearly described, but can only be learned by experience....

6.7 Paracelsus, 'Alchemy, Art of Transformation' [1520s–1530s], trans. Norbert Guterman, in Jolande Jacobi (ed.), *Paracelsus: Selected Writings* (London: Routledge, Kegan and Paul, 1951), pp. 215–23

Nothing has been created as *ultima materia* – in its final state. Everything is at first created in its *prima materia*, its original stuff; whereupon Vulcan comes, and by the art of alchemy develops it into its final substance.... For alchemy means: to carry to its end something that has not yet been completed. To obtain the lead from the ore and to transform it into what it is made for. ...Accordingly, you should understand that alchemy is nothing but the art which makes the impure into the pure through fire.... It can separate the useful from the useless, and transmute it into its final substance and its ultimate essence.

The transmutation of metals is a great mystery of nature. However laborious and difficult this task may be, whatever impediments and obstacles may lie in the way of its accomplishment, this transmutation does not go counter to nature, nor is it incompatible with the order of God, as is falsely asserted by many persons. But the base, impure five metals – that is, copper, tin, lead, iron, and quicksilver – cannot be transmuted into the nobler pure, and perfect metals – namely, into gold and silver – without a *tinctura*, or without the philosopher's stone. [...]

The great virtues that lie hidden in nature would never have been revealed if alchemy had not uncovered them and made them visible. Take a tree, for example; a man sees it in the winter, but he does not know what it is, he does not know what it conceals within itself, until summer comes and discloses the buds, the flowers, the fruit.... Similarly the virtues in things remain concealed to man, unless the alchemists disclose them, as the summer reveals the nature of the tree. – And if the alchemist brings to light that which lies hidden in nature, one must know that those hidden powers are different in each thing – they are different in locusts, different in leaves, different in flowers, and different in ripe and unripe fruits. For all this is so marvellous that in form and qualities the last fruit of a tree is completely unlike the first one.... And each thing has not only one virtue but many, just as a flower has more than one colour, and each colour has in itself the most diverse hues; and yet they constitute a unity, one thing.

Alchemy is a necessary, indispensable art.... It is an art, and Vulcan is its artist. He who is a Vulcan has mastered this art; he who is not a Vulcan can make no headway in it. But to understand this art, one must above all know that God has created all things; and that He has created something out of nothing. This something is a seed, in which the purpose of its use and function is inherent from the beginning. And since all things have been created in an unfinished state, nothing is finished, but Vulcan must bring all things to their completion. Things are created and given into our hands, but not in the ultimate form that is proper to them. For example, wood grows of itself, but does not transform itself into boards or charcoal. Similarly, clay does not of itself become a pot. This is true of everything that grows in nature.

The *quinta essentia* is that which is extracted from a substance – from all plants and from everything which has life – then freed of all impurities and all perishable parts, refined into highest purity and separated from all elements.... The inherency of a thing, its nature, power, virtue, and curative efficacy, without any... foreign admixture... that is the *quinta essentia*. It is a spirit like the life spirit, but with this difference that the *spiritus vitae*, the life spirit, is imperishable, while the spirit of man is perishable.... The *quinta essentia* being the life spirit of things, it can be extracted only from the perceptible, that is to say material, parts, but not from the imperceptible, animated parts of things.... It is endowed with extraordinary powers and perfections, and in it is found a great purity, through which it effects an alteration or cleansing in the body, which is an incomparable marvel.... Thus the *quinta essentia* can cleanse a man's life.... Therefore each disease requires its own *quinta essentia*, although some forms of the *quinta essentia* are said to be useful in all diseases. [...]

But to write more about this mystery is forbidden and further revelation is the prerogative of the divine power. For this art is truly a gift of God. Wherefore not everyone can understand it. For this reason God bestows it upon whom He pleases, and it cannot be wrested from Him by force; for it is His will that He alone shall be honoured in it and that through it His name be praised for ever and ever.

6.8 Paracelsus, (a) 'Seven Defensiones' [1538], (b) 'On the Miners' Sickness' [1533–4], in *Four Treatises of Theophrastus von Hohenheim called Paracelsus*, ed. Henry Sigerist (Baltimore, MD: John Hopkins University Press, 1941), (a) pp. 16–24, (b) pp. 61–4

(a) 'Seven Defensiones'

THE OTHER DEFENCE CONCERNING THE NEW DISEASES AND *Nomina* OF THE ABOVE-MENTIONED DOCTOR THEOPHRASTUS

To defend and protect myself, to shelter myself, in that I describe and depict new diseases never before described, and new *Nomina* never before employed, but given by me: Pay heed therefore why this happens – it is because of new diseases, that I may indicate them. I write of the crazy dance which the common man calls St. Vitus' dance, also of those who kill themselves, also of false maladies which befall through sorcery, as well as of people possessed. It seems unjust to me that these diseases should never have been described by medicine, that they should have been forgotten. But what causes me to do this and brings me to it is the fact that astronomy, which heretofore has never been taken up by physicians, teaches me to recognise such diseases. If the other physicians had been as experienced in astronomy, they would have been explained and discovered completely long before me. Since, however, *Astronomia* is rejected by physicians, these diseases and many others, together with their true causes, can be neither recognized nor understood. Since now the medicine of the other authors does not flow from the spring from which medicine takes its origin, the origin and spring of which I may boast, should I not then have authority to write differently from another writer? To everyone it is given to speak, to advise and to teach, but it is not given to everyone to speak and teach things of strength. For you know that the Gospel too testifies that when Christ taught, He spoke as One who had authority and not as the scribes and hypocrites. Such authority one should respect as proves itself with works, if one is incredulous of the word. Therefore am I well aware that little as a man can describe in detail how a thing is formed, if he has never seen it with his eyes, compared to a man who has seen it with his eyes, so the same judgment will be passed here between those who speak without reason and those who speak with reason. It is not less so with a sick man; he belongs under the physician, and justly all diseases should be known to them. However, what the physician may not know in one, that he should know in another. For thus too were the talents of the apostles distributed and what is given to each, in that he takes pride: what is not given to him is no disgrace to him. For as God desires each man to be, so he remains. The other authors cannot boast of such talents: they rejoice in their term, and what they cannot accomplish through the *Terminum* they say is impossible to cure.

Further, that I defend myself because I write new *Nomina* and new *Recepta*: at this you should not marvel. It happens not because of my simplicity or ignorance; on the contrary, everyone can well realise that any simple scholar can read such *Nomina* as were given by the Ancients, also their prescriptions, from the paper and recognise them. But what drives me away from them, is that the *Nomina* are put together and composed of so many different languages, that we can nevermore get things of this kind thoroughly into our heads. And they themselves do not understand and recognise their own *Nomina*; moreover in the German tongue the *Nomina* are changed from one village to another. And although some people have written *Pandectas*[1] and so on, they have hit upon other things which to believe is not in me. And this for many reasons. That I should wish to put myself in such danger and should willingly enter upon such an uncertified apprenticeship, this my conscience will not do. For in these same authors no chapter is free of lies and great errors, but something is found there that

---

1 In the fourteenth century, for instance, Matthaeus Sylvaticus wrote 'Pandectae Medicinae', an alphabetically arranged reference book.

spoils it all. How then should these authors please me? I seek not *Rhetoricam* or Latin in them, but I seek medicine of which they can give me no account. Thus too with the prescriptions, they say I write them new receipts and introduce a new procedure. As they have told me to my face: I am to use nothing strange, according to the meaning of God's tenth commandment: Thou shalt not covet strange things. Since now they blame and scold me as a transgressor against the tenth commandment, it is necessary for me to discover what is strange or not strange. Namely, that a man enter not by the right door, is strange; that a man take what belongs not to him. For example, that a man claims to be a physician and is it not, that a man doctor with things in which there is no medicine. Should I be blamed for being able to discover these tricks?

Further, that I should write about people possessed is thoroughly distasteful to them, yet I do it not for this reason. Since fasting and prayer drive out evil spirits, I consider it especially to be recommended to the physician first to seek the kingdom of God and afterwards what he needs will be given unto him. If it is given to him through prayer to make the sick healthy, allow that it is a good purgation. If it is given to him through fasting, allow that it is a good *Confortativum*.[2] Tell me one thing: Is medicine only in herbs, wood and stones, and not in words? Then I will tell you what words are. What is it that words cannot do? As the disease is, so also is the medicine: if the disease is entrusted to the herbs, it will be healed by means of herbs. If it is under the stones, it will also be nourished under them; if it is subject to fasting, it must be driven away by fasting. Possession is the great disease. Now as Christ demonstrates the remedy, why should not I investigate the same writing, as to what the prescriptions contain and are in this malady? Heaven makes disease: the physician drives it away again. Now just as heaven has to yield to the physician, so too must the devil yield through the right ordering of medicine. The neoteric and modern physicians carry on as they do, because the loquacious Mesue[3] did not think of such things, nor others whose *Aemuli* they are.

I have been charged too with giving diseases new *Nomina* which no one recognises nor understands. Why do I not adhere to the old Names? How can I use the old *Nomina* when they are not derived from the origin from which the disease rises, but are only Sur-*nomina*, of which no one truly knows whether he rightly calls the disease by that name or not. Since then I find and recognise such uncertain ground, why should I give myself so much trouble on account of the *Nomina*?... I am concerned only with finding out the origin of a disease and its treatment and with relating the name to this....

---

2 i.e. a strengthening medicine.

3 Paracelsus probably refers to a medieval pharmacological work supposedly written by one Mesue.

To instruct you further as to the new diseases which I announce in one way or another: there are yet more causes which constrain to seek new diseases. For instance, the sky has a different action every day, changes daily its constitution. The reason is, it too is growing older. For in the same way, when a child is born it changes according to its age, the further the more different from youth, down to the *Terminum* of death. Now the sky too was a child, it too had its beginning and is predestined to its end, like man, and death is in it and around it. Now as everything changes with age, the works of the same change also. And if there are changes in the works, of what avail to me then is the rod of young children? For this reason I speak of the present monarchies, because of the age of the firmament and the elements.... Thence there comes such a *Pressura gentium*,[4] such, too, as has never been before: and from this there results a medicine which has never before existed. Therefore can the physician not make shift as does he who says: I manage with the books which were written two thousand years ago. There are never again the same *Causae*: things are more biting now, as both philosophies of heaven and the elements sufficiently prove. The above-mentioned *Doctores* of medicine should consider better what they plainly see, that for instance an unlettered peasant heals more than all of them with all their books and red gowns. And if those gentlemen in their red caps were to hear what was the cause, they would sit in a sack full of ashes as did they in Nineveh. Thus I know now that in accordance with this Defence and for the reasons given, I can well describe and give new *Nomina* and new diseases.

The Third Defence
Concerning the Description of the New Receipts

But above and beyond what has been said, the outcry is still greater among the ununderstanding, supposed and fictional physicians who say that the prescriptions which I write are poison, corrosive, and an extraction of all that is evil and poisonous in nature. To such a contention and outcry my first question, if they were clever enough to answer it, would be, whether they knew what was poison and what not, or whether in poison there is no *Mysterium* of nature? For in this point they are lacking in understanding and ignorant of natural forces. For what has God created that is not blessed with a great gift for the good of man? Why then should poison be rejected and despised, if not the poison but nature be sought? I will give you an example that you may understand my intention. Behold the toad, how poisonous indeed and detestable a creature it is: behold also the great *Mysterium* which is in it concerning the pestilence. If then the *Mysterium* should be despised because of the poisonous and detestable character of the toad, what a mockery that would be! Who is it who composed the

---

4 Affliction of the people.

receipt of nature? Was it not God? Why should I despise His *Compositum*, even though what He mixes seems inadequate to me? It is He in Whose hand abideth all wisdom and He knows where to put each *Mysterium*. Why then should I marvel or let myself be frightened because one part is poison, and despise the other part too? Each thing should be used for what it is ordained and we should have no further fear of it. For God Himself is the physician and the medicine. And every physician should acknowledge the strength of God which Christ interprets to us, saying: And though you drink poison, it will not harm you. Now if the poison conquers not but enters without harm when we use it according to nature's ordered way, why then should poison be despised? Who despises poison, knows not what is in the poison. For the *Arcanum* which is in the poison is so blessed, that the poison detracts nothing from it, nor harms it. Not that I would wish to have satisfied you with this verse and paragraph, or to have defended myself sufficiently; rather is it necessary to give you further account, if I am sufficiently to explain poison.

How is it that you see in me that with which you are all filled and rebuke me for a lentil, when melons lie in you? You rebuke me for my prescriptions: consider yours, how they are. First, for instance, with your purging. Where in all your books is a *Purgatio* that is not poison, or serves not death, or can be used without annoyance, if *Dosis* is not given its proper weight? Now pay heed to what this means: it is not too much, nor too little. He who strikes the middle, receives no poison. And even if I used poison, which you cannot prove, but if I did indeed use it and gave its *Dosin*, am I punishable for this, or not? This I desire to make known to all and sundry. You know that Thyriac is made from the snake *Thyro*: why do you not condemn your thyriac also, since the poison of this snake is in it? But since you see that it is useful, and not harmful, you are silent. If then my medicine is found to be not less than thyriac, why should it pay for being new? Why should it not be just as good as an old one? If you wish justly to explain each poison, what is there that is not poison? All things are poison, and nothing is without poison: the *Dosis* alone makes a thing not poison. For example, every food and every drink, if taken beyond its Dose, is poison: the result proves it. I admit also that poison is poison: that it should, however, therefore be rejected, is impossible. Now since nothing exists which is not poison, why do you correct? Only in order that the poison may do no harm. If I too have corrected in like manner, why then do you punish me? You know that *Argentum vivum* is nothing but poison and daily experience proves it. Now you have this in use, you anoint patients with it, much more thickly than a cobbler anoints leather with grease. You fumigate with its cinnabar, you wash with its sublimate and do not wish people to say it is poison; yet it is poison and you introduce such poison into man. And you say it is healthful and good, it is corrected with white lead, just as if it were not poison. Take them to Nuernberg for examination, the *Recepta* that you and I write, and see there who uses poison or not. For you know not the correction of Mercury, nor its *Dosin*; but you anoint with as much as will go in. One thing I must ask you to think out: whether your *Recepta*, which you say are without poison, can cure the *Caducum*,[5] or not, or *Podagram* or *Apoplexiam*; or whether with your sugar of roses you can cure St. Vitus' dance and the *Lunaticos*, or other similar diseases? Indeed you have not done it with that and will still not do it with that. So it must be something else: why then must you be vexed with me when I take what I must and should take, for what it is ordained? I let Him answer for it, Who composed it thus in the creation of heaven and earth. Moreover, as the art is given us of separating two antagonistic things from one another, why then should poison be said from the beginning to be present? Consider all my *Recepta*, whether it is not my first principle that the good be separated from the bad? Is not this separation my correction? Should I not administer and use such a corrected *Arcanum*, since I can find nothing evil in it, and you much less? You object to my Vitriolum in which there is great mystery and more avail in it than in all the apothecaries' boxes. You cannot say it is poison; if you say it is corrosive, tell me in what form? You must make it so, or else it is not corrosive. If it can be made into a corrosive, you can also prepare it as a *Dulcedinem*,[6] for these are both together. As the preparation is, so too is the vitriol; and what every *Simplex* is in itself, is by art made into many beings, into all shapes and forms like food which stands on the table. If man eats it, it becomes human flesh, through a dog dog's flesh, through a cat cat's flesh. Thus is it with medicine: it becomes what you make of it. If it is possible to make evil out of good, it is also possible to make good out of evil. No one should condemn a thing who knows not its transmutation and who knows not what separation does. Though a thing is poison, it may well be turned into non-poison. Take an example from Arsenic, which is one of the chief poisons and a *Drachma* kills any steed: burn it with *Sal nitri* and it is no longer poison. Eating ten pounds is harmless. Behold then what the difference is and what preparation does.

But he who would punish, the same should first learn so that, when he punishes, he does not disgrace himself. I can well see your foolishness and simplicity in the fact, too, that you know not what you say and that one must make considerable allowance for your good-for-nothing tongues. I write new *Recepta*, for the old are useless. And there are new diseases, demanding new *Recepta*. But mark this in all my prescriptions: whatever I may take, I take that in which is the *Arcanum* against the disease which I am fighting. And notice further what I do to it: I separate what is not *Arcanum* from what is *Arcanum* and give to the *Arcanum* its right *Dosin*. Now I know that I have defended my *Recepta* well, and that you scold me concerning them out of your envious hearts, and offer your useless *Recepta*. If you were of good conscience you would give up: but your hearts are full and your mouths run over. I put five

---

5 Epilepsy.
6 Sweetness.

Defensiones in this work: read them through and you will find the reasons why I make the *Recepta* from the same Simples as you denounce as being poison. Why should I pay because I lay the foundation which you cannot see? If you were experienced in the things in which a physician should be experienced, you would change your minds. But this you should remember: that what turns out to the good of man is not poison. Poison is alone what turns out to the harm of man, what is not of service to him but injurious, as your *Recepta* sufficiently testify, where no art is considered, except only pounding, mixing and pouring. Herewith then I have desired to defend and protect myself, that my *Recepta* are administered and applied according to the order of nature, and that you yourselves know not what you say, but use your tongues like a madman, uncomprehending and unthinking.

(b) 'On the Miners' Sickness'

THE SECOND TRACTATE

ON THE ORIGIN AND BIRTH OF THE MINERS' SICKNESS

The first chapter.

The origin and the source of the miners' sickness are contained and described in the second tractate. First experience in these things should be made known, so that the things which give rise to the cough, the gasping, and the lung sickness with all its additions can be recognized and found through examination and clear discernment. The theory of the two sicknesses, of the lung and of the mine, is divided according to the content of this experience. In the same manner that you may comprehend the things which visibly demonstrate that they make the lung sickness, thus they are also to be discovered according to ascertained philosophy, to be present in the influence that was described in the first tractate. You see that fogs grow externally in the chaos between the heaven and the earth and these fogs act in different ways, some of them producing asthma, coughing and shortwindedness. Now this is the experience which teaches us to comprehend that the fog is the cause. Thus because the fog has its origin in the firmament, there is also a fog in the mine, from which the miners' sickness can arise even more severely than from the external one. Now if the cause of this fog should be sought, it can be found that it comes from the sphere Galaxae;[1] those that make it are also in the earth. Now the mineral of such a fog is also a cause, and the recognition of this mineral gives the knowledge of the cure in the same way, just as a knowledge of fire tells us with what it can be extinguished. Thus must all diseases be recognized, whose cure is then possible. For this reason too death is incurable, since the heaven of this constellation has never been found.

---

1 Milky way.

Therefore know further, just as it is to be understood concerning the fog, thus it is also with rain, frost and the like, including also such a winterly cold, from which short-windedness can likewise arise. These things are all to be considered in the mines.

The second chapter.

Now to speak further of the things, which make asthma, such as cold and heat; for instance, a large heated lung, which is cooled with sudden cold, is also attacked by short-windedness, as also through sour beverages and through sweet ones. Now just as such a sudden cooling of the lung, as well as sour and sweet, make short-windedness, thus is this also to be understood in the mines, in that the work produces a heat in the lungs, and the neighboring cold, which penetrates into the chaos, causes a rapid cooling of the lung after the work is finished. Although the cold is not felt, it is still essential in the alant and in the constitution[2] of the earth and its effect is such as if it had been drunk. The same is also to be understood concerning the acid. As you see that a sloe encloses its acid with a skin, thus the acid of the mine is also enclosed by a covering in the earth; and because we make dwellings in the mines we walk in this acid. Now the one acid comes from vitriol, the other from alum, as it can also be understood concerning the sloes and currants. The conjunction can arrange that the things, the acid and the like, are attracted into the chaos of the earth and the lung is eager for them; now it is injured in the same manner as one who has a special desire to eat chalk or another to drink vinegar and there are very many such desires. Thus just like these desires it attracts to itself, now the alum, now the vitriol, now the saltpeter, etc., and when the lung loses this desire, it fares like a sick person, whose desire turns out for the worst. Thus this acid causes hoarseness of the lung, like vinegar or a sour drink, and afterwards it is possible for short-windedness to arise. Just as you have now been instructed concerning the acid, thus should you also understand concerning the sweetness, which lies enclosed, like the sweetness in the currants; and as we walk in the mines, it is the same as if we ate this sweetness. When we eat the thing of the earth, the same thing happens, as if one were to eat currants with the teeth, the only difference being that the sweetness of the mine is ingested in the chaos. And when the desire misleads the lung, this sweetness produces the miners' sickness, the cause of which will be related in different places.

The third chapter.

Now the things are to be dealt with, which permit the external recognition of hoarseness, among which one is fat. If the lung has a desire for fat, it must expect harm from it, which then shows itself. Now there are several

---

2 Character of the earth.

kinds of fat that we eat, from oil, meat or fish, visible or invisible. For this reason several lung sicknesses are found, since some become hoarse from this fat, others from another. Now that the eyes may see and comprehend, nature teaches that fat is also found in the chaos, under the sun, and also in the earth. Now every fat is nothing but sulphur that is divided in different forms and ways. From this it follows now, that the stars show their action in such matter in the same manner as the vapor from the intestine burns after being lit, and it is only an exhalation. Thus the chaos is also furnished with a fixed sulphur. If this sulphur is seized by the lung, it attaches itself to the latter like a resin externally to a tree. And according to the different types and kinds of minerals, different resins arise in the lung; this resin is the complaint and the cause of the miners' disease. Now the cause, why the chaos becomes a resin, which indeed is not its ultimate matter, is that the lung cannot digest it. In the same manner one recognizes, if the stomach digests poorly that it is weak; due to which various things can happen to it, and such things also happen to the lung here.... Since just as a sudden exhalation which strikes into one can turn the lung to resin, thus it is also possible that such a vapor arises from the minerals, a thing which is often seen in the heavens.

6.9 Oswald Croll, *De signaturis internis rerum*, 1608, trans. and ed. P. M. Maxwell-Stuart in *The Occult in Early Modern Europe* (Basingstoke: Macmillan, 1999), pp. 150–1

Correspondences of signatures of the greater and lesser worlds:

Minor world or microcosm	*Greater world or macrocosm*
Physiognomy or the shape of the face	The face of the sky
Chiromantia	Minerals
Pulse	Movement of the sky
Shortness of breath	East wind: West wind
Horror of fevers	Movement of the earth
Lientery, dysentery, diarrhoea	Floods
Torments of colic	Thundering of winds

There are as many types of colic in human beings as there are natures of winds in the world.

Difficulty in curing the pain of nephritis	Flashes of light
Apoplexy	Eclipse, conjunction, or thunder
Wasting disease, tuberculosis, or dryness of the microcosm	Dryness
Epilepsy	Storm

The generation of epilepsy in the lesser world is the same as that of storm and thunder in the greater; and just as a storm changes and undermines the animal senses and intelligence – as is evidenced by the unusual croaking of French hens, the unwonted singing and clamour of other birds and animals, and especially the furious stinging of flies – so it is among epileptics.

In the macrocosm	*In the microcosm*
(1) A change of air happens when a storm arises	(1) A change occurs in the power to reason at the start of an epileptic fit
(2) Dark clouds follow	(2) Darkened eyes, sleep
(3) Wind	(3) Flatulence in the neck and stomach
(4) Cracking and thunder	(4) Cracking of the bladder and shaking of the body
(5) Flashes of lightning and brilliant sparks	(5) Fiery lights in the eyes
(6) Rain	(6) Foaming saliva
(7) Claps of thunder	(7) Compressed spirits batter the limbs under the tight skin
(8) Calm weather	(8) Return to mind and speech
After muddy roads have been soaked with rain, they are dried out again by the sun.	Just so Man returns to himself when the Sun of the microcosm has restored reason, intellectual capacity and the rest of the body's functions to their former state after a required period of rest.
Stone-quarries, which are the bones of the great mother.	Bones in the flesh: for, as the earth is furnished with stones, so has the body been supplied with bones, which are likened to gold, as one might expect of the same nature.
Earth	Flesh
Great rivers	Large veins
The sea	The bladder
Seven metals in the mountains, or seven planets in the heavens	The seven principal members in human beings

Just as flowers on earth are the colours of the stars, so stars in heaven display the meadow of earth.

There is nothing in the world whose property is not also found in Man the Microcosm. For Almighty God has created no creature wiser or more noble than Man, because the character and primary essence of all animals is repeated in him; and since he is the book of all creatures, he fashions, fabricates and transforms himself into the face of all flesh and the intelligence of every creature, like some poetic Proteus.... There is no person so devout and so just but in him lie concealed hidden seeds of the malign stars, which wise people must smother and suppress by daily prayer, lest they grow and manifest themselves.

Chapter Seven
French Science in the Seventeenth Century

7.1 René Descartes, (a) *The World* [1629–33], 1664, (b) *Treatise on Man* [1629–33], 1664, (c) *Discourse on Method*, 1637, (d) *Principles of Philosophy*, 1644, in *The Philosophical Writings of Descartes*, trans. John Cottingham, Robert Stoothoff and Dugald Murdoch, 3 vols, vol. 1 (Cambridge: Cambridge University Press, 1985), (a) pp. 90–7, (b) pp. 99–108, (c) pp. 119–22, (d) pp. 40, 227–38

***(a)* The World**

Many ... things remain for me to explain here, and I would myself be happy to add several arguments to make my opinions more plausible. But in order to make this long discourse less boring for you, I want to clothe part of it in the guise of a fable, in the course of which I hope the truth will not fail to become sufficiently clear, and will be no less pleasing to see than if I were to set it forth wholly naked.

Description of a new world; and the qualities of the matter of which it is composed

For a while, then, allow your thought to wander beyond this world to view another world – a wholly new one which I shall bring into being before your mind in imaginary spaces. The philosophers tell us that such spaces are infinite, and they should certainly be believed, since it is they themselves who invented them. But in order to keep this infinity from hampering and confusing us, let us not try to go right to the end: let us enter it only far enough to lose sight of all the creatures that God made five or six thousand years ago; and after stopping in some definite place, let us suppose that God creates anew so much matter all around us that in whatever direction our imagination may extend, it no longer perceives any place which is empty.

Even though the sea is not infinite, people on some vessel in the middle of it may stretch their view seemingly to infinity; and yet there is more water beyond what they see. Likewise, although our imagination seems able to stretch to infinity, and this new matter is not supposed to be infinite, yet we can suppose that it fills spaces much greater than all those we have imagined. And just to ensure that this supposition contains nothing you might find objectionable, let us not allow our imagination to extend as far as it could; let us intentionally confine it to a determinate space which is no greater, say, than the distance between the earth and the principal stars in the heavens, and let us suppose that the matter which God has created extends indefinitely far beyond in all directions. For it is much more reasonable to prescribe limits to the action of our mind than to the works of God, and we are much better able to do so.

Now since we are taking the liberty of fashioning this matter as we fancy, let us attribute to it, if we may, a nature in which there is absolutely nothing that everyone cannot know as perfectly as possible. To this end, let us expressly suppose that it does not have the form of earth, fire, or air, or any other more specific form, like that of wood, stone, or metal. Let us also suppose that it lacks the qualities of being hot or cold, dry or moist, light or heavy, and of having any taste, smell, sound, colour, light, or other such quality in the nature of which there might be said to be something which is not known clearly by everyone.

On the other hand, let us not also think that this matter is the 'prime matter' of the philosophers, which they have stripped so thoroughly of all its forms and qualities that nothing remains in it which can be clearly understood. Let us rather conceive it as a real, perfectly solid body which uniformly fills the entire length, breadth and depth of this huge space in the midst of which we have brought our mind to rest. Thus, each of its parts always occupies a part of that space which it fits so exactly that it could neither fill a larger one nor squeeze into a smaller; nor could it, while remaining there, allow another body to find a place there.

Let us add that this matter may be divided into as many parts having as many shapes as we can imagine, and that each of its parts is capable of taking on as many motions as we can conceive. Let us suppose, moreover, that God really divides it into many such parts, some larger and some smaller, some of one shape and some of another, however we care to imagine them. It is not that God separates these parts from one another so that there is some void between them: rather, let us regard the differences he creates within this matter as consisting wholly in the diversity of the motions he gives to its parts. From the first instant of their creation, he causes some to start moving in one direction and others in another, some faster and others slower (or even, if you wish, not at all); and he causes them to continue moving thereafter in accordance with the ordinary laws of nature. For God has established these laws in such a marvellous way that even if we suppose he creates nothing beyond what I have mentioned, and sets up no order or proportion within it but composes from it a chaos as confused and muddled as any the poets could describe, the laws of nature are sufficient to cause the parts of this chaos to disentangle themselves and arrange themselves in such good order that they will have the form of a quite perfect world [...].

The laws of nature of this new world

But I do not want to delay any longer telling you by what means nature alone can untangle the confusion of the chaos of which I have spoken, and what the laws are that God has imposed on it.

Note, in the first place, that by 'nature' here I do not mean some goddess or any other sort of imaginary power. Rather, I am using this word to signify matter itself, in so far as I am considering it taken together with all the qualities I have attributed to it, and under the condition that God continues to preserve it in the same way that he created it. For it follows of necessity, from the mere fact that he continues thus to preserve it, that there must be many changes in its parts which cannot, it seems to me, properly be attributed to the action of God (because that action never changes), and which therefore I attribute to nature. The rules by which these changes take place I call the 'laws of nature'.

In order to understand this better, recall that among the qualities of matter, we have supposed that its parts have had various different motions from the moment they were created, and furthermore that they are all in contact with each other on all sides without there being any void between any two of them. From this it follows necessarily that from the time they began to move, they also began to change and diversify their motions by colliding with one another. So if God subsequently preserves them in the same way that he created them, he does not preserve them in the same state. That is to say, with God always acting in the same way and consequently always producing substantially the same effect, there are, as if by accident, many differences in this effect. And it is easy to accept that God, who is, as everyone must know, immutable, always acts in the same way. But without involving myself any further in these metaphysical considerations, I shall set out two or three of the principal rules according to which it must be thought that God causes the nature of this new world to operate. These, I believe, will suffice to acquaint you with all the others.

The first is that each individual part of matter continues always to be in the same state so long as collision with others does not force it to change that state. That is to say, if the part has some size, it will never become smaller unless others divide it; if it is round or square, it will never change that shape unless others force it to; if it is brought to rest in some place, it will never leave that place unless others drive it out; and if it has once begun to move, it will always continue with an equal force until others stop or retard it. [...]

I suppose as a second rule that when one body pushes another it cannot give the other any motion unless it loses as much of its own motion at the same time; nor can it take away any of the other's motion unless its own is increased by as much. [...]

But even if everything our senses ever experienced in the real world seemed manifestly contrary to what is contained in these two rules, the reasoning which has taught them to me seems so strong that I cannot help believing myself obliged to posit them in the new world I am describing to you. For what more firm and solid foundation could one find for establishing a truth, even if one wished to choose it at will, than the very firmness and immutability which is in God?

So it is that these two rules follow manifestly from the mere fact that God is immutable and that, acting always in the same way, he always produces the same effect. For, supposing that God placed a certain quantity of motion in all matter in general at the first instant he created it, we must either admit that he always preserves the same amount of motion in it, or not believe that he always acts in the same way....

I shall add, as a third rule, that when a body is moving, even though its motion for the most part takes place along a curved path and...it can never make any movement which is not in some way circular, yet each of its parts individually tends always to continue moving along a straight line. And so the action of these parts – i.e. the tendency they have to move – is different from their motion. [...]

... [W]hen you swing a stone in a sling, not only does it fly straight out as soon as it leaves the sling, but also while it is in the sling it presses against the middle of it and causes the cord to stretch. This makes it obvious that it always has a tendency to go in a straight line and that it goes in a circle only under constraint.

This rule is based on the same foundation as the other two: it depends solely on God's preserving each thing by a continuous action, and consequently on his preserving it not as it may have been some time earlier but precisely as it is at the very instant that he preserves it. So it is that of all motions, only motion in a straight line is entirely simple and has a nature which may be wholly grasped in an instant. [...]

... [W]e shall, if you please, suppose in addition that God will never perform any miracle in the new world....

(b) Treatise on Man

...First I must describe the body on its own; then the soul, again on its own; and finally I must show how these two natures would have to be joined and united in order to constitute men who resemble us.

I suppose the body to be nothing but a statue or machine made of earth, which God forms with the explicit intention of making it as much as possible like us. Thus God not only gives it externally the colours and shapes of all the parts of our bodies, but also places inside it all the parts required to make it walk, eat, breathe, and indeed to imitate all those of our functions which can be imagined to proceed from matter and to depend solely on the disposition of our organs.

We see clocks, artificial fountains, mills, and other such machines which, although only man-made, have the power to move of their own accord in many different ways. But I am supposing this machine to be made by the hands of God, and so I think you may reasonably think it capable of a greater variety of movements than I could possibly imagine in it, and of exhibiting more artistry than I could possibly ascribe to it. [...]

The parts of the blood which penetrate as far as the brain serve not only to nourish and sustain its substance, but also and primarily to produce in it a certain very fine wind,[1] or rather a very lively and pure flame, which is called the *animal spirits*. For it must be noted that the arteries which carry blood to the brain from the heart, after dividing into countless tiny branches which make up the minute tissues that are stretched like tapestries at the bottom of the cavities of the brain, come together again around a certain little *gland*[2] situated near the middle of the substance of the brain, right at the entrance to its cavities. The arteries in this region have a great many little holes through which the finer parts of the blood can flow into this gland....

Now in the same proportion as the animal spirits enter the cavities of the brain, they pass from there into the pores of its substance, and from these pores into the nerves. And depending on the varying amounts which enter (or merely tend to enter) some nerves more than others, the spirits have the power to change the shape of the muscles in which the nerves are embedded, and by this means to move all the limbs. Similarly you may have observed in the grottos and fountains in the royal gardens that the mere force with which the water is driven as it emerges from its source is sufficient to move various machines, and even to make them play certain instruments or utter certain words depending on the various arrangements of the pipes through which the water is conducted. [...]

Next, to understand how the external objects which strike the sense organs can prompt this machine to move its limbs in numerous different ways, you should consider that the tiny fibres (which, as I have already told you, come from the innermost region of its brain and compose the marrow of the nerves) are so arranged in each part of the machine that serves as the organ of some sense that they can easily be moved by the objects of that sense. And when they are moved, with however little force, they simultaneously pull the parts of the brain from which they come, and thereby open the entrances to certain pores in the internal sur-

---

1 Descartes means 'composed of very small, fast-moving particles'.

2 The pineal gland, which Descartes identified as the seat of the imagination and the ... soul.

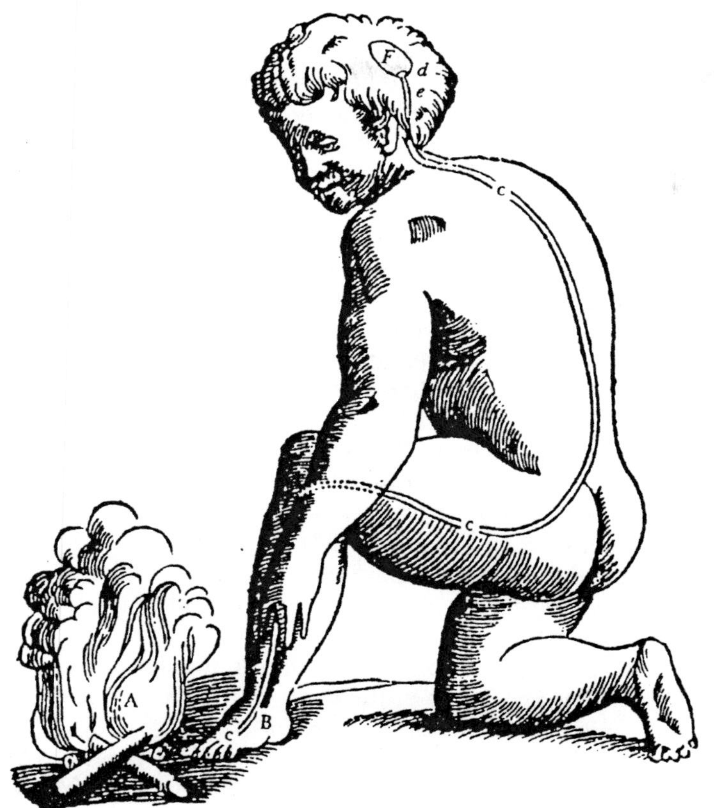

Fig. 1

face of the brain. Through these pores the animal spirits in the cavities of the brain immediately begin to make their way into the nerves and so to the muscles which serve to cause movements in the machine quite similar to those we are naturally prompted to make when our senses are affected in the same way.

Thus, for example [in Fig. 1], if fire A is close to foot B, the tiny parts of this fire (which, as you know, move about very rapidly) have the power also to move the area of skin which they touch. In this way they pull the tiny fibre *cc* which you see attached to it, and simultaneously open the entrance to the pore *de*, located opposite the point where this fibre terminates – just as when you pull one end of a string, you cause a bell hanging at the other end to ring at the same time.

When the entrance to the pore or small tube *de* is opened in this way, the animal spirits from cavity *F* enter and are carried through it – some to muscles which serve to pull the foot away from the fire, some to muscles which turn the eyes and head to look at it, and some to muscles which make the hands move and the whole body turn in order to protect it...

Now I maintain that when God unites a rational soul to this machine... he will place its principal seat in the brain, and will make its nature such that the soul will have different sensations corresponding to the different ways in which the entrances to the pores in the internal surface of the brain are opened by means of the nerves.

Suppose, firstly, that the tiny fibres which make up the marrow of the nerves are pulled with such force that they are broken and separated from the part of the body to which they are joined, with the result that the structure of the whole machine becomes somehow less perfect. Being pulled in this way, the fibres cause a movement in the brain which gives occasion for the soul (whose place of residence must remain constant) to have the sensation of *pain*.

Now suppose the fibres are pulled with a force almost as great as the one just mentioned, but without their being broken or separated from the parts to which they are attached. Then they will cause a movement in the brain which, testifying to the good condition of the other parts of the body, will give the soul occasion to feel a certain bodily pleasure which we call '*titillation*'. This, as you see, is very close to pain in respect of its cause but quite opposite in its effect. [...]

It is time for me to begin to explain how the animal spirits make their way through the cavities and pores of the brain of this machine, and which of the machine's functions depend on these spirits. [...]

In order... to see clearly how ideas are formed of the objects which strike the senses, observe in this diagram [Fig. 2] the tiny fibres 12, 34, 56, and the like, which make up the optic nerve and stretch from the back of the eye at

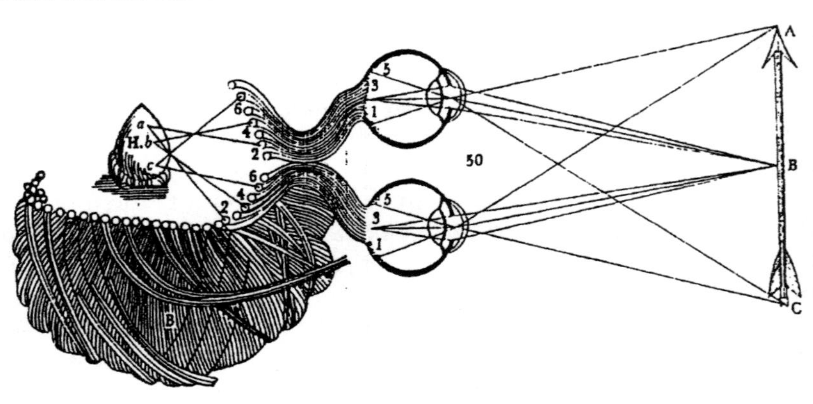

Fig. 2

1, 3, 5 to the internal surface of the brain at 2, 4, 6. Now assume that these fibres are so arranged that if the rays coming, for example, from point A of the object happen to press upon the back of the eye at point 1, they pull the whole of fibre 12 and enlarge the opening of the tiny tube marked 2. In the same way, the rays which come from point B enlarge the opening of the tiny tube 4, and likewise for the others.... [D]epending on the different ways in which the points 1, 3, 5 are pressed by these rays, a figure is traced on the back of the eye corresponding to that of the object ABC. Similarly, it is obvious that, depending on the different ways in which the tiny tubes 2, 4, 6 are opened by the fibres 12, 34, 56, etc., a corresponding figure must also be traced on the internal surface of the brain.

... [J]ust as a figure corresponding to that of the object ABC is traced on the internal surface of the brain according to the different ways in which tubes 2, 4, 6 are opened, likewise that figure is traced on the surface of the gland according to the ways in which the spirits leave from points *a*, *b*, *c*.

And note that by 'figures' I mean not only things which somehow represent the position of the edges and surfaces of objects, but also anything which, as I said above, can give the soul occasion to perceive movement, size, distance, colours, sounds, smells and other such qualities. And I also include anything that can make the soul feel pleasure, pain, hunger, thirst, joy, sadness and other such passions. For it is easy to understand that tube 2, for example, may be opened in different ways – in one way by the action which I said causes sensory perception of the colour red, or of tickling, and in another way by the action which I said causes sensory perception of the colour white, or of pain; and the spirits which leave from point *a* will tend to move towards this tube in a different manner according to differences in its manner of opening, and likewise for the others.

Now among these figures, it is not those imprinted on the external sense organs, or on the internal surface of the brain, which should be taken to be ideas – but only those which are traced in the spirits on the surface of the gland H (*where the seat of the imagination and the 'common' sense is located*). That is to say, it is only the latter figures which should be taken to be the forms or images which the rational soul united to this machine will consider directly when it imagines some object or perceives it by the senses. [...]

... [S]uppose that after the spirits leaving gland H have received the impression of some idea, they pass through tubes 2, 4, 6, and the like, into the pores or gaps lying between the tiny fibres which make up part B of the brain. And suppose that the spirits are strong enough to enlarge these gaps somewhat, and to bend and arrange in various ways any fibres they encounter, according to the various ways in which the spirits are moving and the different openings of the tubes into which they pass. Thus they also trace figures in these gaps, which correspond to those of the objects. At first they do this less easily and perfectly than they do on gland H, but gradually they do it better and better, as their action becomes stronger and lasts longer, or is repeated more often. That is why these figures are no longer so easily erased, and why they are preserved in such a way that the ideas which were previously on the gland can be formed again long afterwards without requiring the presence of the objects to which they correspond. And this is what *memory* consists in. [...]

I should like you to consider, after this, all the functions I have ascribed to this machine – such as the digestion of food, the beating of the heart and arteries, the nourishment and growth of the limbs, respiration, waking and sleeping, the reception by the external sense organs of light, sounds, smells, tastes, heat and other such qualities, the imprinting of the ideas of these qualities in the organ of the 'common' sense and the imagination, the retention or stamping of these ideas in the memory, the internal movements of the appetites and passions, and finally the external movements of all the limbs.... I should like you to consider that these functions follow from the mere arrangement of the machine's organs every bit as naturally as the movements of a clock or other automaton follow from the arrangement of its counter-weights and wheels....

(c) Discourse on Method

[...] When I was younger, my philosophical studies had included some logic, and my mathematical studies some geometrical analysis and algebra. These three arts or sciences, it seemed, ought to contribute something to my plan. But on further examination I observed with regard to logic that syllogisms and most of its other techniques are of less use for learning things than for explaining to others the things one already knows or even, as in the art of Lully, for speaking without judgment about matters of which one is ignorant. And although logic does contain many excellent and true precepts, these are mixed up with so many others which are harmful or superfluous that it is almost as difficult to distinguish them as it is to carve a Diana or a Minerva from an unhewn block of marble. As to the analysis of the ancients and the algebra of the moderns, they cover only highly abstract matters, which seem to have no use. Moreover the former is so closely tied to the examination of figures that it cannot exercise the intellect without greatly tiring the imagination; and the latter is so confined to certain rules and symbols that the end result is a confused and obscure art which encumbers the mind, rather than a science which cultivates it. For this reason I thought I had to seek some other method comprising the advantages of these three subjects but free from their defects. Now a multiplicity of laws often provides an excuse for vices, so that a state is much better governed when it has but few laws which are strictly observed; in the same way, I thought, in place of the large number of rules that make up logic, I would find the following four to be sufficient, provided that I made a strong and unswerving resolution never to fail to observe them.

The first was never to accept anything as true if I did not have evident knowledge of its truth: that is, carefully to avoid precipitate conclusions and preconceptions, and to include nothing more in my judgements than what presented itself to my mind so clearly and so distinctly that I had no occasion to doubt it.

The second, to divide each of the difficulties I examined into as many parts as possible and as may be required in order to resolve them better.

The third, to direct my thoughts in an orderly manner, by beginning with the simplest and most easily known objects in order to ascend little by little, step by step, to knowledge of the most complex, and by supposing some order even among objects that have no natural order of precedence.

And the last, throughout to make enumerations so complete, and reviews so comprehensive, that I could be sure of leaving nothing out.

Those long chains composed of very simple and easy reasonings, which geometers customarily use to arrive at their most difficult demonstrations, had given me occasion to suppose that all the things which can fall under human knowledge are interconnected in the same way. And I thought that, provided we refrain from accepting anything as true which is not, and always keep to the order required for deducing one thing from another, there can be nothing too remote to be reached in the end or too well hidden to be discovered. I had no great difficulty in deciding which things to begin with, for I knew already that it must be with the simplest and most easily known. Reflecting, too, that of all those who have hitherto sought after truth in the sciences, mathematicians alone have been able to find any demonstrations – that is to say, certain and evident reasonings – I had no doubt that I should begin with the very things that they studied. From this, however, the only advantage I hoped to gain was to accustom my mind to nourish itself on truths and not to be satisfied with bad reasoning. Nor did I have any intention of trying to learn all the special sciences commonly called 'mathematics'.[1] For I saw that, despite the diversity of their objects, they agree in considering nothing but the various relations or proportions that hold between these objects. And so I thought it best to examine only such proportions in general, supposing them to hold only between such items as would help me to know them more easily. At the same time I would not restrict them to these items, so that I could apply them the better afterwards to whatever others they might fit. Next I observed that in order to know these proportions I would need sometimes to consider them separately, and sometimes merely to keep them in mind or understand many together. And I thought that in order the better to consider them separately

1 These are subjects with a theoretical basis in mathematics, such as astronomy, music and optics.

I should suppose them to hold between lines, because I did not find anything simpler, nor anything that I could represent more distinctly to my imagination and senses. But in order to keep them in mind or understand several together, I thought it necessary to designate them by the briefest possible symbols. In this way I would take over all that is best in geometrical analysis and in algebra, using the one to correct all the defects of the other.

In fact, I venture to say that by strictly observing the few rules I had chosen, I became very adept at unravelling all the questions which fall under these two sciences. So much so, in fact, that in the two or three months I spent in examining them – beginning with the simplest and most general and using each truth I found as a rule for finding further truths – not only did I solve many problems which I had previously thought very difficult, but also it seemed to me towards the end that even in those cases where I was still in the dark I could determine by what means and to what extent it was possible to find a solution. This claim will not appear too arrogant if you consider that since there is only one truth concerning any matter, whoever discovers this truth knows as much about it as can be known. For example, if a child who has been taught arithmetic does a sum following the rules, he can be sure of having found everything the human mind can discover regarding the sum he was considering. In short, the method which instructs us to follow the correct order, and to enumerate exactly all the relevant factors, contains everything that gives certainty to the rules of arithmetic.

But what pleased me most about this method was that by following it I was sure in every case to use my reason, if not perfectly, at least as well as was in my power. Moreover, as I practised the method I felt my mind gradually become accustomed to conceiving its objects more clearly and distinctly; and since I did not restrict the method to any particular subject-matter, I hoped to apply it as usefully to the problems of the other sciences as I had to those of algebra. Not that I would have dared to try at the outset to examine every problem that might arise, for that would itself have been contrary to the order which the method prescribes. But observing that the principles of these sciences must all be derived from philosophy, in which I had not yet discovered any certain ones, I thought that first of all I had to try to establish some certain principles in philosophy. And since this is the most important task of all, and the one in which precipitate conclusions and preconceptions are most to be feared, I thought that I ought not try to accomplish it until I had reached a more mature age than twenty-three, as I then was, and until I had first spent a long time in preparing myself for it. I had to uproot from my mind all the wrong opinions I had previously accepted, amass a variety of experiences to serve as the subject-matter of my reasonings, and practise constantly my self-prescribed method in order to strengthen myself more and more in its use. [. . .]

(d) Principles of Philosophy

The Principles of Material Things

1. The arguments that lead to the certain knowledge of the existence of material things.
Everyone is quite convinced of the existence of material things. But earlier on we cast doubt on this belief and counted it as one of the preconceived opinions of our childhood. So it is necessary for us to investigate next the arguments by which the existence of material things may be known with certainty. Now, all our sensations undoubtedly come to us from something that is distinct from our mind. For it is not in our power to make ourselves have one sensation rather than another; this is obviously dependent on the thing that is acting on our senses. Admittedly one can raise the question of whether this thing is God or something different from God. But we have sensory awareness of, or rather as a result of sensory stimulation we have a clear and distinct perception of, some kind of matter, which is extended in length, breadth and depth, and has various differently shaped and variously moving parts which give rise to our various sensations of colours, smells, pain and so on. And if God were himself immediately producing in our mind the idea of such extended matter, or even if he were causing the idea to be produced by something which lacked extension, shape and motion, there would be no way of avoiding the conclusion that he should be regarded as a deceiver. For we have a clear understanding of this matter as something that is quite different from God and from ourselves or our mind; and we appear to see clearly that the idea of it comes to us from things located outside ourselves, which it wholly resembles. And we have already noted that it is quite inconsistent with the nature of God that he should be a deceiver. The unavoidable conclusion, then, is that there exists something extended in length, breadth and depth and possessing all the properties which we clearly perceive to belong to an extended thing. And it is this extended thing that we call 'body' or 'matter'. [...]

11. There is no real difference between space and corporeal substance.
It is easy for us to recognize that the extension constituting the nature of a body is exactly the same as that constituting the nature of a space. There is no more difference between them than there is between the nature of a genus or species and the nature of an individual. Suppose we attend to the idea we have of some body, for example a stone, and leave out everything we know to be non-essential to the nature of body: we will first of all exclude hardness, since if the stone is melted or pulverized it will lose its hardness without thereby ceasing to be a body; next we will exclude colour, since we have often seen stones so transparent as to lack colour; next we will exclude heaviness, since although fire is extremely light it is still thought of as being corporeal; and finally we will exclude cold and heat and all other such qualities, either because they are not thought of as being in the stone, or because if they change, the stone is not on that account reckoned to have lost its

bodily nature. After all this, we will see that nothing remains in the idea of the stone except that it is something extended in length, breadth and depth. Yet this is just what is comprised in the idea of a space – not merely a space which is full of bodies, but even a space which is called 'empty'. [...]

13. What is meant by 'external place'.
The terms 'place' and 'space' ... do not signify anything different from the body which is said to be in a place; they merely refer to its size, shape and position relative to other bodies. To determine the position, we have to look at various other bodies which we regard as immobile; and in relation to different bodies we may say that the same thing is both changing and not changing its place at the same time. For example, when a ship is under way, a man sitting on the stern remains in one place relative to the other parts of the ship with respect to which his position is unchanged; but he is constantly changing his place relative to the neighbouring shores, since he is constantly receding from one shore and approaching another. Then again, if we believe the earth moves, and suppose that it advances the same distance from west to east as the ship travels from east to west in the corresponding period of time, we shall again say that the man sitting on the stern is not changing his place; for we are now determining the place by means of certain fixed points in the heavens. [...]

20. The foregoing results also demonstrate the impossibility of atoms.
We also know that it is impossible that there should exist atoms, that is, pieces of matter that are by their very nature indivisible.... For if there were any atoms, then no matter how small we imagined them to be, they would necessarily have to be extended; and hence we could in our thought divide each of them into two or more smaller parts, and hence recognize their divisibility. For anything we can divide in our thought must, for that very reason, be known to be divisible.... Even if we imagine that God has chosen to bring it about that some particle of matter is incapable of being divided into smaller particles, it will still not be correct, strictly speaking, to call this particle indivisible. For, by making it indivisible by any of his creatures, God certainly could not thereby take away his own power of dividing it, since it is quite impossible for him to diminish his own power. ... Hence, strictly speaking, the particle will remain divisible, since it is divisible by its very nature. [...]

22. Similarly, the earth and the heavens are composed of one and the same matter; and there cannot be a plurality of worlds.
It can also easily be gathered from this that celestial matter is no different from terrestrial matter. And even if there were an infinite number of worlds, the matter of which they were composed would have to be identical; hence, there cannot in fact be a plurality of worlds, but only one. For we very clearly understand that the matter whose nature consists simply in its

being an extended substance already occupies absolutely all the imaginable space in which the alleged additional worlds would have to be located; and we cannot find within us an idea of any other sort of matter.

23. *All the variety in matter, all the diversity of its forms, depends on motion.*
The matter existing in the entire universe is thus one and the same, and it is always recognized as matter simply in virtue of its being extended. All the properties which we clearly perceive in it are reducible to its divisibility and consequent mobility in respect of its parts, and its resulting capacity to be affected in all the ways which we perceive as being derivable from the movement of the parts. [...]

25. *What is meant by 'motion' in the strict sense of the term.*
...[M]otion is the transfer of one piece of matter, or one body, from the vicinity of the other bodies which are in immediate contact with it, and which are regarded as being at rest, to the vicinity of other bodies. [...]

33. *How in every case of motion there is a complete circle of bodies moving together.*
... [E]very place is full of bodies.... It follows from this that each body can move only in a complete circle of matter, or ring of bodies which all move together at the same time: a body entering a given place expels another, and the expelled body moves on and expels another, and so on, until the body at the end of the sequence enters the place left by the first body at the precise moment when the first body is leaving it. We can easily understand this in the case of a perfect circle, since we see that no vacuum and no rarefaction or condensation is needed to enable part A of the circle [see Fig. 3] to move towards B, provided that B simultaneously moves towards C, C towards D and D towards A. [...]

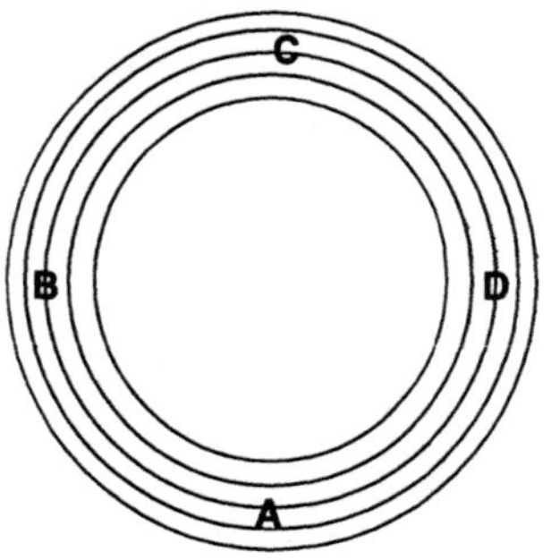

Fig. 3

The Visible Universe

[...]

2. *We must beware of being so presumptuous as to think we understand the ends which God set before himself in creating the world.*

... [I]t would be the height of presumption if we were to imagine that all things were created by God for our benefit alone, or even to suppose that the power of our minds can grasp the ends which he set before himself in creating the universe.

3. *The sense in which it may be said that all things were created for man.*

... [I]t is wholly improbable that all things were in fact made for our benefit, in the sense that they have no other use. And in the study of physics such a supposition would be utterly ridiculous and inept, since there is no doubt that many things exist, or once existed, though they are now here no longer, which have never been seen or thought of by any man, and have never been of any use to anyone.

4. *Experiential phenomena and their use in philosophy.*

... [W]e aim to deduce an account of effects from their causes, not to deduce an account of causes from their effects....

15. *The observed motions of the planets may be explained by various hypotheses.*

When a sailor on the high seas in calm weather looks out from his own ship and sees other ships a long way off changing their mutual positions, he can often be in doubt whether the motion responsible for this change of position should be attributed to this or that ship, or even to his own. In the same way, the paths of the planets, when seen from the earth, are of a kind which makes it impossible for us to know, simply on the basis of the observed motions, what proper motions should be attributed to any given body. And since their paths are very uneven and are very complicated, it is not easy to explain them except by selecting one pattern, among all those which can make their movements intelligible, and supposing the movements to occur in accordance with it. To this end, astronomers have produced three different hypotheses, i.e. suppositions, which are regarded not as being true, but merely as being suitable for explaining the appearances.

16. *Ptolemy's hypothesis does not account for the appearances.*

The first of these hypotheses is that of Ptolemy. Since this is in conflict with many observations made recently (especially the waxing and waning phases of light which are observed on Venus just as they are on the moon), it is now commonly rejected by all philosophers, and hence I will here pass over it.

17. There is no difference between the hypotheses of Copernicus and Tycho, if they are considered simply as hypotheses.

The second hypothesis is that of Copernicus and the third that of Tycho Brahe. These two, considered simply as hypotheses, account for the appearances in the same manner and do not differ greatly, except that the Copernican version is a little simpler and clearer. Indeed, Tycho would have had no occasion to change it, had he not been attempting to unfold the actual truth of things, as opposed to a mere hypothesis.

18. Tycho attributes less motion to the earth than Copernicus, if we go by what he actually says, but in reality he attributes more motion to it.

Copernicus had no hesitation in attributing motion to the earth, but Tycho wished to correct him on this point, regarding it as absurd from the point of view of physics, and in conflict with the common opinion of mankind. But he did not pay sufficient attention to the true nature of motion, and hence, despite his verbal insistence that the earth is at rest, in actual fact he attributed more motion to it than did Copernicus.

19. My denial of the earth's motion is more careful than the Copernican view and more correct than Tycho's view.

The only difference between my position and those of Copernicus and Tycho is that I propose to avoid attributing any motion to the earth, thus keeping closer to the truth than Tycho while at the same time being more careful than Copernicus. I will put forward the hypothesis that seems to be the simplest of all both for understanding the appearances and for investigating their natural causes. And I wish this to be considered simply as a hypothesis or supposition that may be false and not as the real truth. [...]

28. Strictly speaking, the earth does not move, any more than the planets, although they are all carried along by the heaven.

Here we must bear in mind what I said above about the nature of motion, namely that if we use the term 'motion' in the strict sense and in accordance with the truth of things, then motion is simply the transfer of one body from the vicinity of the other bodies which are in immediate contact with it, and which are regarded as being at rest, to the vicinity of other bodies. But it often happens that, in accordance with ordinary usage, any action whereby a body travels from one place to another is called 'motion'; and in this sense it can be said that the same thing moves and does not move at the same time, depending on how we determine its location. It follows from this that in the strict sense there is no motion occurring in the case of the earth or even the other planets, since they are not transferred from the vicinity of those parts of the heaven with which they are in immediate contact, in so far as these parts are considered as being at rest. Such a transfer would require them to move away from all these parts at the same time, which does not occur....

29. *No motion should be attributed to the earth even if 'motion' is taken in the loose sense, in accordance with ordinary usage; but in this sense it is correct to say that the other planets move.*

[...] And if...we appear to attribute motion to the earth, it should be remembered that this is an improper way of speaking – rather like the way in which we may sometimes say that passengers asleep on a ferry 'move' from Calais to Dover, because the ship takes them there.

30. *All the planets are carried round the sun by the heaven.*

Let us thus put aside all worries regarding the earth's motion, and suppose that the whole of the celestial matter in which the planets are located turns continuously like a vortex with the sun at its centre. Further, let us suppose that the parts of the vortex which are nearer the sun move more swiftly than the more distant parts, and that all the planets (including the earth) always stay surrounded by the same parts of celestial matter. This single supposition enables us to understand all the observed movements of the planets with great ease, without invoking any machinery. In a river there are various places where the water twists around on itself and forms a whirlpool. If there is flotsam on the water we see it carried around with the whirlpool, and in some cases we see it also rotating about its own centre; further, the bits which are nearer the centre of the whirlpool complete a revolution more quickly; and finally, although such flotsam always has a circular motion, it scarcely ever describes a perfect circle but undergoes some longitudinal and latitudinal deviations. We can without any difficulty imagine all this happening in the same way in the case of the planets, and this single account explains all the planetary movements that we observe. [...]

43. *If a cause allows all the phenomena to be clearly deduced from it, then it is virtually impossible that it should not be true.*

Suppose, then, that we use only principles which we see to be utterly evident, and that all our subsequent deductions follow by mathematical reasoning: if it turns out that the results of such deductions agree accurately with all natural phenomena, we would seem to be doing God an injustice if we suspected that the causal explanations discovered in this way were false. For this would imply that God had endowed us with such an imperfect nature that even the proper use of our powers of reasoning allowed us to go wrong.

44. *Nevertheless, I want the causes that I shall set out here to be regarded simply as hypotheses.*

When philosophizing about such important matters, however, it would seem to be excessively arrogant for us to assert that we have discovered the exact truth where others have failed; and so I should prefer to leave this claim on one side, and put forward everything that I am about to write simply as a hypothesis which is perhaps far from the truth, so as to leave everyone free to make up his own mind. And if it is thought that the hypothesis is false, I shall think I have achieved something sufficiently worthwhile if everything deduced from it agrees with our observations; for if this is so, we shall see that our hypothesis yields just as much practical benefit for our lives as we would have derived from knowledge of the actual truth....

45. I shall even make some assumptions which are agreed to be false.
Indeed, in order to provide a better explanation for the things found in nature, I shall take my investigation of their causes right back to a time before the period when I believe that the causes actually came into existence. For there is no doubt that the world was created right from the start with all the perfection which it now has. The sun and earth and moon and stars thus existed in the beginning, and, what is more, the earth contained not just the seeds of plants but the plants themselves; and Adam and Eve were not born as babies but were created as fully grown people. This is the doctrine of the Christian faith, and our natural reason convinces us that it was so. For if we consider the infinite power of God, we cannot think that he ever created anything that was not wholly perfect of its kind. Nevertheless, if we want to understand the nature of plants or of men, it is much better to consider how they can gradually grow from seeds than to consider how they were created by God at the very beginning of the world. Thus we may be able to think up certain very simple and easily known principles which can serve, as it were, as the seeds from which we can demonstrate that the stars, the earth and indeed everything we observe in this visible world could have sprung. For although we know for sure that they never did arise in this way, we shall be able to provide a much better explanation of their nature by this method than if we merely described them as they now are or as we believe them to have been created. And since I believe I have in fact found such principles, I shall give a brief account of them here.

46. The assumptions that I am making here in order to give an explanation of all phenomena.
From what has already been said we have established that all the bodies in the universe are composed of one and the same matter, which is divisible into indefinitely many parts, and is in fact divided into a large number of parts which move in different directions and have a sort of circular motion; moreover, the same quantity of motion is always preserved in the universe. However, we cannot determine by reason alone how big these pieces of matter are, or how fast they move, or what kinds of circle they describe. Since there are countless different configurations which God might have instituted here, experience alone must teach us which configurations he actually selected in preference to the rest. We are thus free to make any assumption on these matters with the sole proviso that all the consequences of our assumption must agree with our experience. So, if we may, we will

suppose that the matter of which the visible world is composed was originally divided by God into particles which were approximately equal, and of a size which was moderate, or intermediate when compared with those that now make up the heavens and stars. We will also suppose that the total amount of motion they possessed was equal to that now found in the universe; and that their motions were of two kinds, each of equal force. First, they moved individually and separately about their own centres, so as to form a fluid body such as we take the heavens to be; and secondly, they moved together in groups around certain other equidistant points corresponding to the present centres of the fixed stars, and also around other rather more numerous points equalling the number of the planets and the comets, ... so as to make up as many different vortices as there are now heavenly bodies in the universe.

47. The falsity of these suppositions does not prevent the consequences deduced from them being true and certain.

These few assumptions seem to me to be sufficient to serve as the causes or principles from which all the effects observed in our universe would arise in accordance with the laws of nature set out above. And I do not think it is possible to think up any alternative principles for explaining the real world that are simpler, or easier to understand, or even more probable. It may be possible to start from primeval chaos as described by the poets, i.e. a total confusion in all parts of the universe, and deduce from it, in accordance with the laws of nature, the precise organization now to be found in things; and I once undertook to provide such an explanation. But confusion seems less in accordance with the supreme perfection of God the creator of all things than proportion or order; and it is not possible for us to have such a distinct perception of it. What is more, no proportion or order is simpler or easier to know than that characterized by complete equality in every respect. This is why I am supposing at this point that all the particles of matter were initially equal in respect both of their size and their motion; and I am allowing no inequality in the universe beyond that which exists in the position of the fixed stars, which is so clearly apparent to anyone looking at the night sky that it is quite impossible to deny it. In fact it makes very little difference what initial suppositions are made, since all subsequent change must occur in accordance with the laws of nature. And there is scarcely any supposition that does not allow the same effects (albeit more laboriously) to be deduced in accordance with the same laws of nature. For by the operation of these laws matter must successively assume all the forms of which it is capable; and, if we consider these forms in order, we will eventually be able to arrive at the form which characterizes the universe in its present state. Hence in this connection we need not fear that any error can arise from a false supposition.

48. How all the particles of celestial matter become spherical.

49. There must be other more subtle matter, more tiny particles, around these spherical particles to fill all the space in that area.

50. The particles of this more subtle matter can be very easily divided.

51. And they move very quickly.

52. There are three elements of this visible world.
We have...two very different kinds of matter which can be said to be the first two elements of this visible universe. The first element is made up of matter which is so violently agitated that when it meets other bodies it is divided into particles of indefinite smallness...The second is composed of matter divided into spherical particles which are still very minute when compared with those that we can see with our eyes, but which have a definite fixed quantity and can be divided into other much smaller particles. The third element, which we shall discover a little later on, consists of particles which are much bulkier or have shapes less suited for motion. From these elements, as we shall show, all the bodies of this visible universe are composed. The sun and fixed stars are composed of the first element, the heavens from the second, and the earth with the planets and comets from the third.... [...]

55. What light is.
It is a law of nature that all bodies moving in a circle move away from the centre of their motion in so far as they can. I shall now explain as carefully as I can the force by means of which the globules of the second element ...strive to move away from their centres of motion; for the nature of light consists in this alone, as will be shown below, and there are many other matters which depend on knowledge of this point.

56. The striving after motion in inanimate things, and how it should be understood.
When I say that the globules of the second element 'strive' to move away from the centres around which they revolve, it should not be thought that I am implying that they have some thought from which this striving proceeds. I mean merely that they are positioned and pushed into motion in such a way that they will in fact travel in that direction, unless they are prevented by some other cause.

57. How the same body can be said to strive to move in different directions at the same time.
Often many different causes act simultaneously on the same body, and one may hinder the effect of another. So, depending on the causes we are considering, we may say that the body is tending or striving to move in different directions at the same time. For example, the stone A [see Fig. 5].... [...]

205. That it is morally certain that all the things of this world are such as it has been demonstrated here that they can be.

[...] Suppose for example that someone wants to read a letter written in Latin but encoded so that the letters of the alphabet do not have their proper value, and he guesses that the letter B should be read whenever A appears, and C when B appears, i.e. that each letter should be replaced by the one immediately following it. If, by using this key, he can make up Latin words from the letters, he will be in no doubt that the true meaning of the letter is contained in these words. It is true that his knowledge is based merely on a conjecture, and it is conceivable that the writer did not replace the original letters with their immediate successors in the alphabet, but with others, thus encoding quite a different message; but this possibility is so unlikely (especially if the message contains many words) that it does not seem credible. Now if people look at all the many properties relating to magnetism, fire and the fabric of the entire world, which I have deduced in this book from just a few principles, then, even if they think that my assumption of these principles was arbitrary and groundless, they will still perhaps acknowledge that it would hardly have been possible for so many items to fit into a coherent pattern if the original principles had been false.

206. Indeed, my explanations possess more than moral certainty.

Besides, there are some matters, even in relation to the things in nature, which we regard as absolutely... certain. Absolute certainty arises when we believe that it is wholly impossible that something should be otherwise

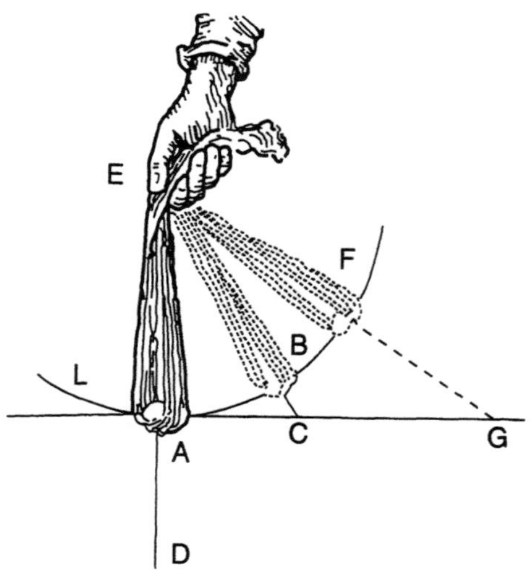

Fig. 5

than we judge it to be. This certainty is based on a metaphysical foundation, namely that God is supremely good and in no way a deceiver, and hence that the faculty which he gave us for distinguishing truth from falsehood cannot lead us into error, so long as we are using it properly and are thereby perceiving something distinctly. Mathematical demonstrations have this kind of certainty, as does the knowledge that material things exist; and the same goes for all evident reasoning about material things. And perhaps even these results of mine will be allowed into the class of absolute certainties, if people consider how they have been deduced in an unbroken chain from the first and simplest principles of human knowledge. Their certainty will be especially appreciated if it is properly understood that we can have no sensory awareness of external objects unless these objects produce some local motion in our nerves; and that the fixed stars, owing to their enormous distance from us, cannot produce such motion unless there is also some motion occurring both in them and also throughout the entire intervening part of the heavens. Once this is accepted, then it seems that all the other phenomena, or at least the general features of the universe and the earth which I have described, can hardly be intelligibly explained except in the way I have suggested.

207. I submit all my views to the authority of the Church.
Nevertheless, mindful of my own weakness, I make no firm pronouncements, but submit all these opinions to the authority of the Catholic Church and the judgement of those wiser than myself. And I would not wish anyone to believe anything except what he is convinced of by evident and irrefutable reasoning.

7.2 Blaise Pascal, *Story of the Great Experiment on the Equilibrium of Fluids*, 1648, trans. I. Spiers and A. Spiers from *The Physical Treatises of Pascal* (New York: Octagon Books, 1973), pp. 97–112

The weather on Saturday last, the nineteenth of this month, was very unsettled. At about five o'clock in the morning, however, it seemed sufficiently clear; and since the summit of the Puy de Dôme was then visible, I decided to go there to make the attempt. To that end I notified several people of standing in this town of Clermont, who had asked me to let them know when I would make the ascent. Of this company some were clerics, others laymen. Among the clerics was the Very Revd. Father Bannier, one of the Minim Fathers of this city, who has on several occasions been 'Corrector' (that is, Father Superior), and the Monsieur Mosnier, Canon of the Cathedral Church of this city; among the laymen were Messieurs La Ville and Begon, councillors to the Court of Aids, and Monsieur La Porte, a doctor of medicine, practising here. All these men are very able, not only in the practice of their professions, but also in every field of intellectual interest. It was a delight to have them with me in this fine work.

On that day, therefore, at eight o'clock in the morning, we started off all together for the garden of the Minim Fathers, which is almost the lowest spot in the town, and there began the experiment in this manner.

First, I poured into a vessel six pounds of quicksilver which I had rectified during the three days preceding; and having taken glass tubes of the same size, each four feet long and hermetically sealed at one end but open at the other, I placed them in the same vessel and carried out with each of them the usual vacuum experiment. Then, having set them up side by side without lifting them out of the vessel, I found that the quicksilver left in each of them stood at the same level, which was twenty-six inches and three and a half lines above the surface of the quicksilver in the vessel. I repeated this experiment twice at this same spot, in the same tubes, with the same quicksilver, and in the same vessel; and found in each case that the quicksilver in the two tubes stood at the same horizontal level, and at the same height as in the first trial.

That done, I fixed one of the tubes permanently in its vessel for continuous experiment. I marked on the glass the height of the quicksilver, and leaving that tube where it stood, I requested Revd. Father Chastin, one of the brothers of the house, a man as pious as he is capable, and one who reasons very well upon these matters, to be so good as to observe from time to time all day any changes that might occur. With the other tube and a portion of the same quicksilver, I then proceeded with all these gentlemen to the top of the Puy de Dôme, some 500 fathoms above the Convent. There, after I had made the same experiments in the same way that I had made them at the Minims, we found that there remained in the tube, a height of only twenty-three inches and two lines of quicksilver; whereas in the same tube, at the Minims we had found a height of twenty-six inches and three and a half lines. Thus between the heights of the quicksilver in the two experiments there proved to be a difference of three inches one line and a half. We were so carried away with wonder and delight, and our surprise was so great that we wished, for our own satisfaction, to repeat the experiment. So I carried it out with the greatest care five times more at different points on the summit of the mountain, once in the shelter of the little chapel that stands there, once in the open, once shielded from the wind, once in the wind, once in fine weather, once in the rain and fog which visited us occasionally. Each time I most carefully rid the tube of air; and in all these experiments we invariably found the same height of quicksilver. This was twenty-three inches and two lines, which yields the same discrepancy of three inches, one line and a half in comparison with the twenty-six inches, three lines and a half which had been found at the Minims. This satisfied us fully.

Later, on the way down at a spot called Lafon de l'Arbre, far above the Minims but much farther below the top of the mountain, I repeated the same experiment, still with the same tube, the same quicksilver, and the same vessel, and there found that the height of the quicksilver left in the tube was twenty-five inches. I repeated it a second time at the same spot; and Monsieur Mosnier, one of those previously mentioned, having the curiosity to perform it himself, then did so again, at the same spot. All these experiments yielded the same height of twenty-five inches, which is one inch, three lines and a half less than that which we had found at the Minims, and one inch and ten lines more than we had just found at the top of the Puy de Dôme. It increased our satisfaction not a little to observe in this way that the height of the quicksilver diminished with the altitude of the site.

On my return to the Minims I found that the [quicksilver in the] vessel I had left there in continuous operation was at the same height at which I had left it, that is, at twenty-six inches, three lines and a half; and the Revd. Father Chastin, who had remained there as observer, reported to us that no change had occurred during the whole day, although the weather had been very unsettled, now clear and still, now rainy, now very foggy, and now windy.

Here I repeated the experiment with the tube I had carried to the Puy de Dôme, but in the vessel in which the tube used for the continuous experiment was standing. I found that the quicksilver was at the same level in both tubes and exactly at the height of twenty-six inches, three lines and a half, at which it had stood that morning in this same tube, and as it had stood all day in the tube used for the continuous experiment.

I repeated it again a last time, not only in the same tube I had used on the Puy de Dôme, but also with the same quicksilver and in the same vessel that I had carried up the mountain; and again I found the quicksilver at the same height of twenty-six inches, three lines and a half which I had observed in the morning, and thus finally verified the certainty of our results.

Chapter Eight
Science in Seventeenth-Century England

8.1 William Gilbert, *De Magnete*, 1600, trans. P. Fleury Mottelay (New York: Dover edition, 1958), pp. 23–5, 327, 333–5

The magnetic poles may be found in every loadstone, whether strong and powerful (male, as the term was in antiquity) or faint, weak, and female; whether its shape is due to design or to chance, and whether it be long, or flat, or four-square, or three-cornered, or polished; whether it be rough, broken-off, or unpolished: the loadstone ever has and ever shows its poles. But inasmuch as the spherical form, which, too, is the most perfect, agrees best with the earth, which is a globe, and also is the form best suited for experimental uses, therefore we purpose to give our principal demonstrations with the aid of a globe-shaped loadstone, as being the best and the most fitting. Take then a strong loadstone, solid, of convenient size, uniform, hard, without flaw; on a lathe, such as is used in turning crystals and some precious stones, or on any like instrument (as the nature and toughness of the stone may require, for often it is worked only with difficulty), give the loadstone the form of a ball. The stone thus prepared is a true homogeneous offspring of the earth and is of the same shape, having got from art the orbicular form that nature in the beginning gave to the earth, the common mother; and it is a natural little body endowed with a multitude of properties whereby many abstruse and unheeded truths of philosophy, hid in deplorable darkness, may be more readily brought to the knowledge of mankind. To this round stone we give the name *Μικρόγη* (microge) or Terrella (earthkin, little earth).[1]

---

1 Sir Kenelm Digby, 'A Treatise of Bodies' (London, 1645), Chap. XX, p. 225.

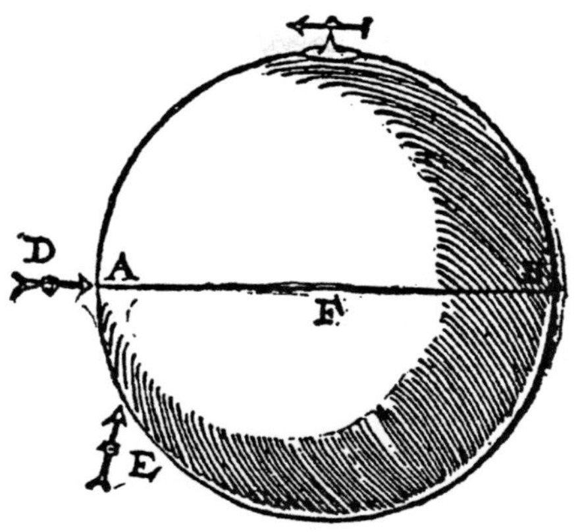

To find, then, poles answering to the earth's poles, take in your hand the round stone, and lay on it a needle or a piece of iron wire: the ends of the wire move round their middle point, and suddenly come to a standstill. Now, with ochre or with chalk, mark where the wire lies still and sticks. Then move the middle or centre of the wire to another spot, and so to a third and a fourth, always marking the stone along the length of the wire where it stands still: the lines so marked will exhibit meridian circles, or circles like meridians on the stone or terrella; and manifestly they will all come together at the poles of the stone. The circles being continued in this way, the poles appear, both the north and the south, and betwixt these, midway, we may draw a large circle for an equator, as is done by the astronomer in the heavens and on his spheres and by the geographer on the terrestrial globe; for the line so drawn on this our terrella is also of much utility in our demonstrations and our magnetic experiments. Poles are also found in the round stone, in a versorium, in a piece of iron touched with a loadstone and resting on a needle or point (attached at its base to the terrella), so that it can freely revolve, as in the figure [above]. [...]

On top of the stone *AB* is set the versorium in such a way that its pointer may remain in equilibrium: mark with chalk the direction of the pointer when at rest. Then move the instrument to another spot and again mark the direction in which the pointer looks; repeat this many times at many different points and you will, from the convergence of the lines of direction, find one pole at the point *A*, the other at *B*. A pointer also indicates the true pole if brought near to the stone, for it eagerly faces the stone at right angles, and seeks the pole itself direct and turns on its axis in a right line toward the centre of the stone. Thus the pointer *D* regards *A* and *F*, the pole and the centre, but the pointer *E* looks not straight either toward the pole *A* or the centre *F*. A bit of fine iron wire as long as a barley-corn is laid on the stone and is moved over the zones and the surface of the stone till it stands perpendicularly erect; for at the poles, whether N. or S., it stands erect; but the farther it is from the poles (towards the equator) the more it inclines. The poles thus found, you are to mark with a sharp file or a gimlet. [...]

[...] We infer, not with mere probability, but with certainty, the diurnal rotations of the earth; for nature ever acts with fewer rather than with many means; and because it is more accordant to reason that the one small body, the earth, should make a daily revolution than that the whole universe should be whirled around it. I pass by the earth's other movements, for here we treat only of the diurnal rotation, whereby it turns to the sun and produces the natural day (of twenty-four hours) which we call *nycthemeron*. And, indeed, nature would seem to have given a motion quite in harmony with the shape of the earth, for the earth being a globe, it is far easier and far more fitting that it should revolve on its natural poles, than that the whole universe, whose bounds we know not nor can know, should be whirled round; easier and more fitting than that there should be fashioned a sphere of the *primum mobile* – a thing not received by the ancients, and which even Aristotle never thought of or admitted as existing beyond the sphere of the fixed stars; finally, which the holy Scriptures do not recognize, as neither do they recognize a revolution of the whole firmament. [...]

The earth therefore rotates, and by a certain law of necessity, and by an energy that is innate, manifest, conspicuous, revolves in a circle toward the sun; through this motion it shares in the solar energies and influences; and its verticity holds it in this motion lest it stray into every region of the sky. The sun (chief inciter of action in nature), as he causes the planets to advance in their courses, so, too, doth bring about this revolution of the globe by sending forth the energies of his spheres – his light being effused.

And were not the earth to revolve with diurnal rotation, the sun would ever hang with its constant light over a given part, and, by long tarrying there, would scorch the earth, reduce it to powder, and dissipate its substance, and the uppermost surface of earth would receive grievous hurt: nothing of good would spring from earth, there would be no vegetation; it could not give life to the animate creation, and man would perish. In other parts all would be horror, and all things frozen stiff with intense cold: hence all its eminences would be hard, barren, inaccessible, sunk in everlasting shadow and unending night. And as the earth herself cannot endure so pitiable and so horrid a state of things on either side, with her astral magnetic mind she moves in a circle, to the end there may be, by unceasing change of light, a perpetual vicissitude, heat and cold, rise and decline, day and night, morn and even, noonday and deep night. So the earth seeks and seeks the sun again, turns from him, follows him, by her wondrous magnetical energy.

And not only from the sun would ill impend, were the earth to stand still and be deprived of the benefit of his rays; from the moon also great dangers would threaten. For we see how the ocean swells and comes to flood under certain positions of the moon. But if by the daily rotation of the earth the moon did not quickly pass, the sea would rise unduly at some parts and many coasts would be overwhelmed by mighty tides. Lest the earth, then, should in divers ways perish and be destroyed, she rotates in virtue of her magnetic and primary energy. And such are the movements in the rest of the planets, the motion and light of other bodies especially urging. For the moon also turns round during its menstrual circuit that it may on all its parts successively receive the sun's light, which it enjoys, with which it is refreshed like the earth itself; nor could the moon without grave ill and sure destruction stand the unceasing incidence of the light on one of its sides only.

Thus each of the moving globes has circular motion, either in a great circular orbit or on its own axis or in both ways. But that all the fixed stars, and the planets, and all the higher heavens, still revolve simply for the earth's sake is for the mind of a philosopher a ridiculous supposition. The earth then revolves, and not the whole heavens; and this movement brings growth and decay, gives occasion for the generation of animated things, and arouses the internal heat to productiveness. Hence does matter vegetate to receive forms, and from this primary revolution of the earth natural bodies have prime incitation and original act. The motion of the whole earth, therefore, is primary, astral, circular about its poles, whose verticity rises on both sides from the plane of the equator, and the energy is infused into the opposite ends, so that the globe by a definite rotation might move to the good, sun and stars inciting. But the simple right-downward motion assumed by the Peripatetics is the movement of weight, of coacervation, of separated parts, in the ratio of their matter, by right lines toward the earth's centre, these tending to the centre by the shortest route. The motions of separate magnetical parts of the earth are, besides that of coacervation, those of coition, revolution, and direction of the parts to the whole, into harmony and agreement of the form.

8.2 William Harvey, *De Motu Cordis et Sanguinis*, 1628 (London: Dent, Everyman edition, 1952), ch. viii pp. 55–7, ch. ix pp. 58–9

OF THE QUANTITY OF BLOOD PASSING THROUGH THE HEART FROM THE VEINS TO THE ARTERIES; AND OF THE CIRCULAR MOTION OF THE BLOOD

Thus far I have spoken of the passage of the blood from the veins into the arteries, and of the manner in which it is transmitted and distributed by the action of the heart; points to which some, moved either by the authority of Galen or Columbus, or the reasonings of others, will give in their adhesion. But what remains to be said upon the quantity and source of the blood which thus passes, is of so novel and unheard-of character, that I not only fear injury to myself from the envy of a few, but I tremble lest I have mankind at large for my enemies, so much doth wont and custom, that become as another nature, and doctrine once sown and that hath struck deep root, and respect for antiquity influence all men: Still the die is cast, and my trust is in my love of truth, and the candour that inheres in cultivated minds. And sooth to say, when I surveyed my mass of evidence, whether derived from vivisections, and my various reflections on them, or from the ventricles of the heart and the vessels that enter into and issue from them, the symmetry and size of these conduits, – for nature doing nothing in vain, would never have given them so large a relative size without a purpose, – or from the arrangement and intimate structure of the valves in particular, and of the other parts of the heart in general, with many things besides, I frequently and seriously bethought me, and long revolved in my mind, what might be the quantity of blood which was transmitted, in how short a time its passage might be effected, and the like; and not finding it possible that this could be supplied by the juices of the ingested aliment without the veins on the one hand becoming drained, and the arteries on the other getting ruptured through the excessive charge of blood, unless the blood should somehow find its way from the arteries into the veins, and so return to the right side of the heart; I began to think whether there might not be a A MOTION, AS IT WERE, IN A CIRCLE. Now this I afterwards found to be true; and I finally saw that the blood, forced by the action of the left ventricle into the arteries, was distributed to the body at large, and its several parts, in the same manner as it is sent through the lungs, impelled by the right ventricle into the pulmonary artery, and that it then passed through the veins and along the vena cava, and so round to the left ventricle in the manner already indicated. Which motion we may be allowed to call circular, in the same way as Aristotle says that the air and the rain emulate the circular motion of the superior bodies; for the moist earth, warmed by the sun, evaporates; the vapours drawn upwards are condensed, and descending in the form of rain, moisten the earth again; and by this arrangement are generations of living things produced; and in like manner too are tempests and meteors engendered by the circular motion, and by the approach and recession of the sun.

And so, in all likelihood, does it come to pass in the body, through the motion of the blood; the various parts are nourished, cherished, quickened by the warmer, more perfect, vaporous, spirituous, and, as I may say, alimentive blood; which, on the contrary, in contact with these parts becomes cooled, coagulated, and, so to speak, effete; whence it returns to its sovereign the heart, as if to its source, or to the inmost home of the body, there to recover its state of excellence or perfection. Here it resumes its due fluidity and receives an infusion of natural heat – powerful, fervid, a kind of treasury of life, and is impregnated with spirits, and it might be said with balsam; and thence it is again dispersed; and all this depends on the motion and action of the heart.

The heart, consequently, is the beginning of life; the sun of the microcosm, even as the sun in his turn might well be designated the heart of the world; for it is the heart by whose virtue and pulse the blood is moved, perfected, made apt to nourish, and is preserved from corruption and coagulation; it is the household divinity which, discharging its function, nourishes, cherishes, quickens the whole body, and is indeed the foundation of life, the source of all action. [...]

THAT THERE IS A CIRCULATION OF THE BLOOD IS CONFIRMED FROM THE FIRST PROPOSITION

But lest any one should say that we give them words only, and make mere specious assertions without any foundation, and desire to innovate without sufficient cause, three points present themselves for confirmation, which being stated, I conceive that the truth I contend for will follow necessarily, and appear as a thing obvious to all. First, – the blood is incessantly transmitted by the action of the heart from the vena cava to the arteries in such quantity, that it cannot be supplied from the ingesta, and in such wise that the whole mass must very quickly pass through the organ; Second, – the blood under the influence of the arterial pulse enters and is impelled in a continuous, equable, and incessant stream through every part and member of the body, in much larger quantity than were sufficient for nutrition, or than the whole mass of fluids could supply; Third, – the veins in like manner return this blood incessantly to the heart from all parts and members of the body. These points proved, I conceive it will be manifest that the blood circulates, revolves, propelled and then returning, from the heart to the extremities, from the extremities to the heart, and thus that it performs a kind of circular motion.

Let us assume either arbitrarily or from experiment, the quantity of blood which the left ventricle of the heart will contain when distended to be, say two ounces, three ounces, one ounce and a half – in the dead body I have found it to hold upwards of two ounces. Let us assume further, how much less the heart will hold in the contracted than in the dilated state; and how much blood it will project into the aorta upon each contraction, – and all the world allows that with the systole something is always projected, a necessary consequence demonstrated in the third chapter, and obvious from the structure of the valves; and let us suppose as approaching the truth that the fourth, or fifth, or sixth, or even but the eighth part of its charge is thrown into the artery at each contraction; this would give either half an ounce, or three drachms, or one drachm of blood as propelled by the heart at each pulse into the aorta; which quantity, by reason of the valves at the root of the vessel, can by no means return into the ventricle. Now, in the course of half an hour, the heart will have made more than one thousand beats, in some as many as two, three, or even four thousand. Multiplying the number of drachms propelled by the number of pulses, we shall have either one thousand half-ounces, or one thousand times three drachms, or a like proportional quantity of blood, according to the amount which we assume as propelled with each stroke of the heart, sent from this organ into the artery; a larger quantity in every case than is contained in the whole body! In the same way, in the sheep or dog, say that but a single scruple of blood passes with each stroke of the heart, in one half-hour we should have one thousand scruples, or about three pounds and a half of blood injected into the aorta; but the body of neither animal contains above four pounds of blood, a fact which I have myself ascertained in the case of the sheep.

Upon this supposition, therefore, assumed merely as a ground for reasoning, we see the whole mass of blood passing through the heart, from the veins to the arteries, and in like manner through the lungs.

8.3 Francis Bacon, (a) Epistle Dedicatory, (b) Preface to *The New Organon*, 1620, (c) Aphorisms, in *The Philosophical Works of Francis Bacon*, ed. J. Robertson (from text of Ellis, Spedding and Heath, *Works*, 1857–9) (Freeport, NY: Books For Libraries Press, 1905, repr. 1970), (a) p. 242, (b) p. 256, (c) pp. 259–65

(a) Epistle Dedicatory

EPISTLE DEDICATORY

TO OUR MOST GRACIOUS AND MIGHTY PRINCE AND LORD JAMES, BY THE GRACE OF GOD OF GREAT BRITAIN, FRANCE, AND IRELAND KING, DEFENDER OF THE FAITH, ETC.

Most Gracious and Mighty King,

YOUR Majesty may perhaps accuse me of larceny, having stolen from your affairs so much time as was required for this work. I know not what to say for myself. For of time there can be no restitution, unless it be that what has been abstracted from your business may perhaps go to the memory of your name and the honour of your age; if these things are indeed worth anything. Certainly they are quite new; totally new in their very kind: and yet they are copied from a very ancient model; even the world itself and the nature of things and of the mind. And to say truth, I am wont for my own part to regard this work as a child of time rather than of wit; the only wonder being that the first notion of the thing, and such great suspicions concerning matters long established, should have come into any man's mind. All the rest follows readily enough. And no doubt there is something of accident (as we call it) and luck as well in what men think as in what they do or say. But for this accident which I speak of, I wish that if there be any good in what I have to offer, it may be ascribed to the infinite mercy and goodness of God, and to the felicity of your Majesty's times; to which as I have been an honest and affectionate servant in my life, so after my death I may yet perhaps, through the kindling of this new light in the darkness of philosophy, be the means of making this age famous to posterity; and surely to the times of the wisest and most learned of kings belongs of right the regeneration and restoration of the sciences. Lastly, I have a request to make – a request no way unworthy of your Majesty, and which especially concerns the work in hand; namely, that you who resemble Solomon in so many things – in the gravity of your judgments, in the peacefulness of your reign, in the largeness of your heart, in the noble variety of the books which you have composed – would further follow his example in taking order for the collecting and perfecting of a Natural and Experimental History true and severe (unincumbered with literature and book-learning), such as philosophy may be built upon, – such, in fact, as I shall in its proper place describe: that so at length, after the lapse of so many ages, philosophy and the sciences may no longer float in air, but rest on the solid foundation of experience of every kind, and the same well examined and weighed. I have provided the machine, but the stuff must be gathered from the facts of nature. May God Almighty long preserve your Majesty!

Your Majesty's
Most bounden and devoted
Servant,
FRANCIS VERULAM,
CHANCELLOR.

(b) Preface to *The New Organon, 1620*

THE SECOND PART OF THE WORK, WHICH IS CALLED
THE NEW ORGANON;
OR, TRUE DIRECTIONS CONCERNING THE INTERPRETATION OF NATURE.

[AUTHOR'S] PREFACE.

THOSE who have taken upon them to lay down the law of nature as a thing already searched out and understood, whether they have spoken in simple assurance or professional affectation, have therein done philosophy and the sciences great injury. For as they have been successful in inducing belief, so they have been effective in quenching and stopping inquiry; and have done more harm by spoiling and putting an end to other men's efforts than good by their own. Those on the other hand who have taken a contrary course, and asserted that absolutely nothing can be known – whether it were from hatred of the ancient sophists, or from uncertainty and fluctuation of mind, or even from a kind of fulness of learning, that they fell upon this opinion, – have certainly advanced reasons for it that are not to be despised; but yet they have neither started from true principles nor rested in the just conclusion, zeal and affectation having carried them much too far. The more ancient of the Greeks (whose writings are lost) took up with better judgment a position between these two extremes, – between the presumption of pronouncing on everything, and the despair of comprehending anything; and though frequently and bitterly complaining of the difficulty of inquiry and the obscurity of things, and like impatient horses champing the bit, they did not the less follow up their object and engage with Nature; thinking (it seems) that this very question, – viz. whether or not anything can be known, – was to be settled not by arguing, but by trying. And yet they too, trusting entirely to the force of their understanding, applied no rule, but made everything turn upon hard thinking and perpetual working and exercise of the mind.

Now my method, though hard to practise, is easy to explain; and it is this. I propose to establish progressive stages of certainty. The evidence of the sense, helped and guarded by a certain process of correction, I retain. But the mental operation which follows the act of sense I for the most part reject; and instead of it I open and lay out a new and certain path for the mind to proceed in, starting directly from the simple sensuous perception. The necessity of this was felt no doubt by those who attributed so much importance to Logic; showing thereby that they were in search of helps for the understanding, and had no confidence in the native and spontaneous process of the mind. But this remedy comes too late to do any good, when the mind is already, through the daily intercourse and conversation of life, occupied with unsound doctrines and beset on all sides by vain imaginations. And therefore that art of Logic, coming (as I said) too late to the rescue, and no way able to set matters right again, has had the effect of fixing errors rather than disclosing truth. There remains but one course for the recovery of a sound and healthy condition, – namely, that the entire work of the understanding be commenced afresh, and the mind itself be from the very outset not left to take its own course, but guided at every step: and the business be done as if by machinery. Certainly if in things mechanical men had set to work with their naked hands, without help or force of instruments, just as in things intellectual they have set to work with little else than the naked forces of the understanding, very small would the matters have been which, even with their best efforts applied in conjunction, they could have attempted or accomplished....

(c) Aphorisms

APHORISMS CONCERNING THE INTERPRETATION OF NATURE AND THE KINGDOM OF MAN.

APHORISM

I.

MAN, being the servant and interpreter of Nature, can do and understand so much and so much only as he has observed in fact or in thought of the course of nature: beyond this he neither knows anything nor can do anything.

II.

Neither the naked hand nor the understanding left to itself can effect much. It is by instruments and helps that the work is done, which are as much wanted for the understanding as for the hand. And as the instruments of the hand either give motion or guide it, so the instruments of the mind supply either suggestions for the understanding or cautions.

III.

Human knowledge and human power meet in one; for where the cause is not known the effect cannot be produced. Nature to be commanded must be obeyed; and that which in contemplation is as the cause is in operation as the rule. [...]

VIII.

Moreover the works already known are due to chance and experiment rather than to sciences; for the sciences we now possess are merely systems for the nice ordering and setting forth of things already invented; not methods of invention or directions for new work.

IX.

The cause and root of nearly all evils in the sciences is this – that while we falsely admire and extol the powers of the human mind we neglect to seek for its true helps.

X.

The subtlety of nature is greater many times over than the subtlety of the senses and understanding; so that all those specious meditations, speculations, and glosses in which men indulge are quite from the purpose, only there is no one by to observe it.

XI.

As the sciences which we now have do not help us in finding out new works, so neither does the logic which we now have help us in finding out new sciences.

XII.

The logic now in use serves rather to fix and give stability to the errors which have their foundation in commonly received notions than to help the search after truth. So it does more harm than good.

XIII.

The syllogism is not applied to the first principles of sciences, and is applied in vain to intermediate axioms; being no match for the subtlety of nature. It commands assent therefore to the proposition, but does not take hold of the thing.

XIV.

The syllogism consists of propositions, propositions consist of words, words are symbols of notions. Therefore if the notions themselves (which is the root of the matter) are confused and over-hastily abstracted from the facts, there can be no firmness in the superstructure. Our only hope therefore lies in a true induction.

XV.

There is no soundness in our notions whether logical or physical. Substance, Quality, Action, Passion, Essence itself, are not sound notions: much less are Heavy, Light, Dense, Rare, Moist, Dry, Generation, Corruption, Attraction, Repulsion, Element, Matter, Form, and the like; but all are fantastical and ill defined. [. . .]

XVIII.

The discoveries which have hitherto been made in the sciences are such as lie close to vulgar notions, scarcely beneath the surface. In order to penetrate into the inner and further recesses of nature, it is necessary that both notions and axioms be derived from things by a more sure and guarded way; and that a method of intellectual operation be introduced altogether better and more certain.

XIX.

There are and can be only two ways of searching into and discovering truth. The one flies from the senses and particulars to the most general axioms, and from these principles, the truth of which it takes for settled and immoveable, proceeds to judgment and to the discovery of middle axioms. And this way is now in fashion. The other derives axioms from the senses and particulars, rising by a gradual and unbroken ascent, so that it arrives at the most general axioms last of all. This is the true way, but as yet untried.

XX.

The understanding left to itself takes the same course (namely, the former) which it takes in accordance with logical order. For the mind longs to spring up to positions of higher generality, that it may find rest there; and so after a little while wearies of experiment. But this evil is increased by logic, because of the order and solemnity of its disputations.

XXI.

The understanding left to itself, in a sober, patient, and grave mind, especially if it be not hindered by received doctrines, tries a little that other way, which is the right one, but with little progress; since the understanding, unless directed and assisted, is a thing unequal, and quite unfit to contend with the obscurity of things. [. . .]

XXXI.

It is idle to expect any great advancement in science from the superinducing and engrafting of new things upon old. We must begin anew from the very foundations, unless we would revolve for ever in a circle with mean and contemptible progress.

XXXII.

The honour of the ancient authors, and indeed of all, remains untouched; since the comparison I challenge is not of wits or faculties, but of ways and methods, and the part I take upon myself is not that of a judge, but of a guide.

XXXIII.

This must be plainly avowed: no judgment can be rightly formed either of my method or of the discoveries to which it leads, by means of anticipations (that is to say, of the reasoning which is now in use); since I cannot be called on to abide by the sentence of a tribunal which is itself on its trial.

XXXIV.

Even to deliver and explain what I bring forward is no easy matter; for things in themselves new will yet be apprehended with reference to what is old. [...]

XXXVI.

One method of delivery alone remains to us; which is simply this: we must lead men to the particulars themselves, and their series and order; while men on their side must force themselves for awhile to lay their notions by and begin to familiarise themselves with facts.

XXXVII.

The doctrine of those who have denied that certainty could be attained at all, has some agreement with my way of proceeding at the first setting out; but they end in being infinitely separated and opposed. For the holders of that doctrine assert simply that nothing can be known; I also assert that not much can be known in nature by the way which is now in use. But then they go on to destroy the authority of the senses and understanding; whereas I proceed to devise and supply helps for the same.

XXXVIII.

The Idols and false notions which are now in possession of the human understanding, and have taken deep root therein, not only so beset men's minds that truth can hardly find entrance, but even after entrance obtained, they will again in the very instauration of the sciences meet and trouble us, unless men being forewarned of the danger fortify themselves as far as may be against their assaults.

XXXIX.

There are four classes of Idols which beset men's minds. To these for distinction's sake I have assigned names, – calling the first class *Idols of the Tribe*; the second, *Idols of the Cave*; the third, *Idols of the Market-place*; the fourth, *Idols of the Theatre*.

XL.

The formation of ideas and axioms by true induction is no doubt the proper remedy to be applied for the keeping off and clearing away of idols. To point them out, however, is of great use; for the doctrine of Idols is to the Interpretation of Nature what the doctrine of the refutation of Sophisms is to common Logic.

XLI.

The Idols of the Tribe have their foundation in human nature itself, and in the tribe or race of men. For it is a false assertion that the sense of man is the measure of things. On the contrary, all perceptions as well of the sense as of the mind are according to the measure of the individual and not according to the measure of the universe. And the human understanding is like a false mirror, which, receiving rays irregularly, distorts and discolours the nature of things by mingling its own nature with it.

XLII.

The Idols of the Cave are the Idols of the individual man. For every one (besides the errors common to human nature in general) has a cave or den of his own, which refracts and discolours the light of nature; owing either to his own proper and peculiar nature; or to his education and conversation with others; or to the reading of books, and the authority of those whom he esteems and admires; or to the differences of impressions, accordingly as they take place in a mind preoccupied and predisposed or in a mind indifferent and settled; or the like. So that the spirit of man (according as it is meted out to different individuals) is in fact a thing variable and full of perturbation, and governed as it were by chance. Whence it was well observed by Heraclitus that men look for sciences in their own lesser worlds, and not in the greater or common world.

XLIII.

There are also Idols formed by the intercourse and association of men with each other, which I call Idols of the Market-place, on account of the commerce and consort of men there. For it is by discourse that men associate; and words are imposed according to the apprehension of the vulgar. And therefore the ill and unfit choice of words wonderfully obstructs the understanding. Nor do the definitions or explanations wherewith in some things learned men are wont to guard and defend themselves, by any means set the matter right. But words plainly force and overrule the understanding, and throw all into confusion, and lead men away into numberless empty controversies and idle fancies.

XLIV.

Lastly, there are Idols which have immigrated into men's minds from the various dogmas of philosophies, and also from wrong laws of demonstration. These I call Idols of the Theatre; because in my judgment all the received systems are but so many stage-plays, representing worlds of their own creation after an unreal and scenic fashion. Nor is it only of the systems now in vogue, or only of the ancient sects and philosophies, that I speak; for many more plays of the same kind may yet be composed and in like artificial manner set forth; seeing that errors the most widely different have nevertheless causes for the most part alike. Neither again do I mean this only of entire systems, but also of many principles and axioms in science, which by tradition, credulity, and negligence have come to be received....

8.4 Robert Hooke, Preface, *Micrographia*, 1665 (New York: Dover Publications, 1961)

IT *is the great prerogative of Mankind above other Creatures, that we are not only able to* behold *the works of Nature, or barely to* sustein *our lives by them, but we have also the power of* considering, comparing, altering, assisting, *and* improving *them to various uses. And as this is the peculiar priviledge of humane Nature in general, so is it capable of being so far advanced by the helps of Art, and Experience, as to*

By the Council of the ROYAL SOCIETY of *London* for Improving of Natural Knowledge.

Ordered, *That the Book written by* Robert Hooke, *M.A. Fellow of this Society, Entituled*, Micrographia, or ſome Phyſiological Deſcriptions of Minute Bodies, made by Magnifying Glaſſes, with Obſervations and Inquiries thereupon, *Be printed by* John Martyn, *and* James Alleſtry, *Printers to the ſaid Society.*

Novem. 23.
1664.

BROUNCKER. *P. R. S.*

MICROGRAPHIA:

OR SOME

Physiological Descriptions

OF

MINUTE BODIES

MADE BY

MAGNIFYING GLASSES.

WITH

OBSERVATIONS and INQUIRIES thereupon.

By *R. HOOKE*, Fellow of the ROYAL SOCIETY.

Non possis oculo quantum contendere Linceus,
Non tamen idcirco contemnas Lippus inungi. Horat. Ep. lib. 1.

LONDON, Printed by *Jo. Martyn*, and *Ja. Allestry*, Printers to the ROYAL SOCIETY, and are to be sold at their Shop at the *Bell* in S. *Paul's* Church-yard. M DC LX V.

make some Men excel others in their Observations, and Deductions, almost as much as they do Beasts. By the addition of such artificial Instruments *and* methods, *there may be, in some manner, a reparation made for the mischiefs, and imperfection, mankind has drawn upon it self, by negligence, and intemperance, and a wilful and superstitious deserting the Prescripts and Rules of Nature, whereby every man, both from a deriv'd corruption, innate and born with him, and from his breeding and converse with men, is very subject to slip into all sorts of errors.*

The only way which now remains for us to recover some degree of those former perfections, seems to be, by rectifying the operations of the Sense, *the* Memory, *and* Reason, *since upon the evidence, the* strength, *the* integrity, *and the* right correspondence *of all these, all the light, by which our actions are to be guided, is to be renewed, and all our command over things is to be establisht.*

It is therefore most worthy of our consideration, to recollect their several defects, that so we may the better understand how to supply them, and by what assistances we may inlarge *their power, and* secure *them in performing their particular duties.*

As for the actions of our Senses, *we cannot but observe them to be in many particulars much outdone by those of other Creatures, and when at best, to be far short of the perfection they seem capable of: And these infirmities of the Senses arise from a double cause, either from the* disproportion of the Object to the Organ, *whereby an infinite number of things can never enter into them, or else from* error in the Perception, *that many things, which come within their reach, are not received in a right manner.* [...]

The next care to be taken, in respect of the Senses, is a supplying of their infirmities with Instruments, *and, as it were, the adding of* artificial Organs *to the* natural; *this in one of them has been of late years accomplisht with prodigious benefit to all sorts of useful knowledge, by the invention of Optical Glasses. By the means of* Telescopes, *there is nothing so* far distant *but may be represented to our view; and by the help of* Microscopes, *there is nothing so* small, *as to escape our inquiry; hence there is a new visible World discovered to the understanding. By this means the Heavens are open'd, and a vast number of new Stars, and new Motions, and new Productions appear in them, to which all the antient Astronomers were utterly Strangers. By this the Earth it self, which lyes so neer us, under our feet, shews quite a new thing to us, and in every* little particle *of its matter, we now behold almost as great a variety of Creatures, as we were able before to reckon up in the whole* Universe *it self.*

It seems not improbable, but that by these helps the subtilty of the composition of Bodies, the structure of their parts, the various texture of their matter, the instruments and manner of their inward motions, and all the other possible appearances of things, may come to be more fully discovered; all which the antient Peripateticks *were content to comprehend in two general and (unless further explain'd) useless words of* Matter *and* Form. *From whence there may arise many admirable advantages, towards the increase of the* Operative, *and the* Mechanick *Knowledge, to which this Age seems so much inclined, because we may perhaps be inabled to discern all the secret workings of Nature, almost in the same manner as we do those that are the productions of Art, and are manag'd by Wheels, and Engines, and Springs, that were devised by humane Wit.* [. . .]

And I beg my Reader, to let me take the boldness to assure him, that in this present condition of knowledge, a man so qualified, as I have indeavoured to be, only with resolution, and integrity, and plain intentions of imploying his Senses *aright, may venture to compare the reality and the usefulness of his services, towards the true Philosophy, with those of other men, that are of much stronger, and more acute* speculations, *that shall not make use of the same method by the Senses.*

The truth is, the Science of Nature has been already too long made only a work of the Brain *and the* Fancy: *It is now high time that it should return to the plainness and soundness of* Observations *on* material *and* obvious *things. It is said of great Empires, That* the best way to preserve them from decay, is to bring them back to the first Principles, and Arts, on which they did begin. *The same is undoubtedly true in Philosophy, that by wandring far away into* invisible Notions, *has almost quite destroy'd it self, and it can never be recovered, or continued, but by returning into the same* sensible paths, *in which it did at first proceed.* [. . .]

The Indeavours of Skilful men have been most conversant about the assistance of the Eye, and many noble Productions have followed upon it; and from hence we may conclude, that there is a way open'd for advancing the operations, not only of all the other Senses, but even of the Eye it self; that which has been already done ought not to content us, but rather to incourage us to proceed further, and to attempt greater things in the same and different wayes.

'Tis not unlikely, but that there may be yet invented several other helps for the eye, as much exceeding those already found, as those do the bare eye, such as by which we may perhaps be able to discover living Creatures *in the Moon, or other Planets, the* figures *of the compounding Particles of matter, and the particular* Schematisms *and* Textures *of Bodies.*

And as Glasses *have highly promoted our* seeing, *so 'tis not improbable, but that there may be found many* Mechanical Inventions *to improve our other Senses, of* hearing, smelling, tasting, touching. [. . .]

'Tis not improbable also, but that the sense of feeling *may be highly improv'd, for that being a sense that judges of the more* gross *and* robust motions *of the* Particles *of* Bodies, *seems capable of being improv'd and assisted very many wayes. Thus for the distinguishing of* Heat *and* Cold, *the* Weather-glass *and* Thermometer, *which I have describ'd in this following Treatise, do exceedingly perfect it; by each of which the least variations of heat or cold, which the most Acute sense is not able to distinguish, are manifested. This is oftentimes further promoted also by the help of* Burning-glasses, *and the like, which collect and unite the radiating heat. Thus the* roughness *and* smoothness *of a Body is made much more sensible by the help of a* Microscope, *then by the most* tender *and* delicate Hand. *Perhaps, a Physitian might, by several other* tangible *proprieties, discover the constitution of a Body as well as by the* Pulse. [. . .]

8.5 Robert Boyle, *Some Considerations Touching the Usefulnesse of Experimental Natural Philosophy, The Second Tome*, 1671, in M. Hunter and E. Davis (eds), *The Works of Robert Boyle*, vol. 6 (London: Pickering & Chatto, 1999), pp. 396–400 (vols 1–14 1999–2000)

[...] If it be ask'd why I did not forbeare to make use of some Practises of tradesmen and other known, and perhaps seemingly triviall, Experiments. These things may be replyed,

1. That since on divers occasions it was requisite, that my discourse should tend rather to convince than barely to inform my reader, it was proper, that I should imploy at least Some instances, whose truth was generally enough known, or easy to be known (by making inquiry among Artificers) even by such as out of lasiness, or want of Skill, or accommodation cannot conveniently make themselves the tryals.

2. But yet, I have taken care, that these should not be the only, nor yet the most numerous instances, I make use of: it being in this Tome, as well as in my other Physiologicall writings, my main businesse, to take all just Occasions to contribute as much, as without indiscretion I can, to the history of Nature and Arts.

3. As to the Practices and observations of Tradesmen, the two considerations already alledged, may both of them be extended to the giving of an account of the mention I make of them. Of the truth of divers of the Experiments I alledge of theirs, one may be easily satisfyed by inquiring of Artificers about it, and the particular or more circumstantial accounts I give of some of their experiments, I was induc'd to set down by my desire to contribute toward an experimental History. For I have found by long and unwelcome experience, that very few Tradesmen will and can give a man a clear and full account of their own Practices; partly out of Envy, partly out of want of skill to deliver a relation intelligibley enough, and partly (to which I may add *chiefly*) because they omitt generally, to express either at all or at least clearly some important circumstance, which because long use hath made very familiar to them, they presume also to be known to others: and yet the omission of such circumstances, doth often render the Accounts they give of such practices, so darke and so defective, that, if their experiments be any thing intricate or difficult (for if they be Simple and easy, they are not so liable to produce mistakes) I seldom thinke my self sure of their truth, and that I sufficiently comprehend them, till I have either tryed them at home, or caused the Artificers to make them in my presence.

They that have given themselves the trouble of endeavouring to make the experiments of Tradesmen, to be met with in the writings of Cardan, Weckar, and Baptista Porta for instance, and have thereby discovered (what is not usually obvious upon a transient reading) how lamely and darkely, (not to add unintelligibley) severall things are written, will probably afford me their Assent, having found upon tryal the instructions of such learned and ingenious men, to be often obscure and insufficient for practice.

But here I must give the reader notice, that as Mechanical Artes for the most part advance from time to time towards perfection, so the Practices of Artificers may vary in differing times, as well as in differing places, as I have often had occasion to observe. And therefore I would neither have him condemn other writers or Relators, for delivering accounts of the experiments of Craftsmen differing from those I have given, nor condemn me, for having contented my self to set down such Practices faithfully, as I learn't them from the best Artificers (especially those of London) I had opportunity to converse with.

But here perhaps it will be demanded by way of objection, whether I doe not injure Tradesmen by discovering so plainly those things, which our Laws call the Mysteries of their Arts. To a question, that may perhaps by some be clamorously pressed, not only upon me, but much more upon Some ingenious men of our Nation, who pens have bin more bold than mine in disclosing Craftesmens Secrets, 'twill be requisite to return severall things by way of answer, but that such Readers as are not troubled with the Scruple, may not be so with the Apology, they will find this printed in another character, so that, if they please, they may pass it over unread.

First then, It may be represented, that I never divulge *all* the Secrets and practices necessary to the exercise of any one Trade, contenting my self to deliver here and there upon occasion some few particular Experiments, that make for my present purpose: So that, for much more than I allow my self to doe, I can plead the example, not only of other writers, that have published Books to teach the whole Mystery of this, or that trade, as the Priest *Antonio Neri* hath diligently done in his Italian *Arte Vetraria*, and some English, as well as forreign, Virtuosi have done on other Subjects; But also some of the Artificers themselves, as the famous Gold-smith and Jeweller *Benvennio Cellini* in his much esteemed Italian Tracts of the Lapidaries and Goldsmiths Trades. Thus also the famous Mineralist *Georgius Agricola* published in Latin a whole Volume of the more practical part of Mineralogie, wherein he largely and particularly describes Experiments, tooles, and other things that belong to the Callings of *Mein men*. To which I might add divers other Treatises, some of them French, others Italian, (which, though I could not procure them, I have seene among curious collections of books) that have bin published about Severall Artes by the Artificers themselves. And 'tis notorious, that in English, as well as in divers forrein languages, we have Books of the Artes of Gunnery, Distillation, Painting, Gardening &c. divulg'd by persons, that Professed those Callings.

Secondly, it is not the Custome of Tradesmen to buy Books, especially such as are not intended for such Readers, and treat (for the most part) of things either beyond their reach or wherein they seem not likely to be concerned; And as for Gentlemen and Scholars, though some of them may, to satisfy their curiosity, make a few tryalls, yet their doing so will scarce in the least be prejudicial to Tradesmen. Since (to omitt other Arguments) it will not be worth while for a Virtuoso to be at the charge and trouble of buying tooles, and procuring other necessary accommodations to sell a few productions of his skill, though he should not scruple to descend to such a Practice. For if he make but a small number of Experiments, their effects will cost him more than the like may be bought for, of those that make them in great Quantities, and whom their trade obligeth to be sollicitous to buy their instruments and materialls at the best hand, and sell them to the best profit. Besides that most of the workes of Artificers, are chiefly recommended to the more curious sort of buyers by a certain politenesse, and other ornaments (comprised by many under the name of *Finishing*:) which require either an instructed and dexterous hand, or at least some little peculiar directions, which I did not allwayes thinke my self oblig'd to mention, in a treatise designed to assist my friend to become a Philosopher, not a tradesman, and publish'd to help the Reader to gain knowledge not to get mony.

Thirdly, to publish an Experiment or two, or in some cases a much greater number belonging to a Trade, is not sufficient to rob a Tradesman of his Profession. For, besides that most trades consist of Severall parts, and are each of them made up of divers Practices (that commonly are more than a few) Those numerous Mechanicall Arts, that are called handicrafts, require (as their very name argueth a Manuall dexterity) not to be learnt from Bookes, but to be obtaind by imitation and use. And to these considerations I shall add this more important one, that Mechanicall professions are wont to be as it were made up of two parts, which, for distinction sake, I take leave to call the *Art* and the *Craft*; by the former whereof I mean the skill of making such or such things, which are the genuine Productions of the Art, (as when a Taylor maketh a suit, or a cloak,) and by the latter I mean the result of those informations and Experiments, by which the Artificer learns to make the utmost profit, that he can, of the Productions of his Art. And this Oeconomical Prudence is a thing very distinct from the Art it self, and yet is often the most beneficial thing to the Artificer, informing him how to chuse his materialls and estimate their goodnesse and worth, in what places, and at what times, the best and cheapest are to be had, where, and when, and to what persons the things may be most profitably vented. In short, the Craft is that which teacheth him how both to buy his materialls and tooles, and to sell what he makes with them to the most advantage.

Fourthly, it may often prove more advantageous than prejudiciall to Tradesmen themselves, that many of their practices should be known to Experimentall Philosophers....

Yet I shall now represent, that though some little inconvenience may happen to some Tradesmen by the disclosing some of their Experiments to practicall Naturalists, yet that may be more than compensated, partly, by what may be contributed to the perfecting of such experiments themselves, and, partly by the diffused Knowledge and sagacity of Philosophers, and by those new Inventions, which may probably be expected from such persons, especially if they be furnished with Variety of hints from the practices already in use. For these Inventions of ingenious heads doe, when once grown into request, set many Mechanical hands a worke, and supply Tradesmen with new meanes of getting a livelyhood or even inriching themselves. As to the discipline subordinated to the pure Mathematicks, this is very Evident, for those speculative Sciences have (though not Immediately) produced their trades that make Quadrants, Sectors, Astrolabes, Globes, Maps, Lutes, Vialls, Organs, and other Geometrical, Astronomical, Geographical, and Musical instruments; and not to instance those many Trades, that subsist by making such things as Mechanicians, proceeding upon Geometrical Propositions, have bin the Authors of; we know that whether the excellent Galileo was or was not the *first* finder out of Telescopes, yet he improv'd them so much, and by his discoveries in the heavens, did so recommend their usefullnesse to the curious, that many Artificers in divers parts of Europe have thought fit to take up the Trade of making prospective glasses. And since his death, severall others have had profitable worke laid out for them, by the newer directions of some English Gentlemen, deeply skill'd in Dioptricks, and happy at Mechanical contrivances; in so much that now we have severall shops, that furnish not only our own Virtuosi, but those of forrein Countryes with excellent Microscopes and Telescopes, of which latter sort I lately bought one (but I confesse the only one that the maker of it, or any man, that I hear of, hath perfected of that bignesse) which is of threescore foot in length, and which the Ingenious Artist that made it, Mr Reeves, prized constantly at no lesse than an hundred pounds (English mony) I know not, whether or no I should add, that possibly some particular experiments of mine have not bin hitherto unprofitable to severall Tradesmen: But this I may safely affirm, that a great deal of mony hath bin gained by Tradesmen, both in England and elswhere upon the account of the scarlet Dye, invented in our time by Cornelius Drebbell, who was not bred a Dyer nor other Tradesman. And that we dayly see the shops of clockmakers and watchmakers more and more furnished with these usefull instruments, *Pendulum Clocks*, as they are now called, which, but very few years agoe, were brought into request, by that most ingenious Gentleman, who discovered the new Planet about Saturn.

I have handled the Subject of the foregoing Arguments much more particularly, than I would have done, had not my pen bin draw'n on, by a Hope that the things I have represented may furnish Apologies to many inquisitive men, who may be thereby enbolden'd to carry Philosophical materialls from the shops to the Scholes, and divulge the experiments of Artificers, both to the improvement of trades themselves, and to the *great* inriching of the History of Artes and Nature. [...]

8.6 Robert Boyle, *Of the Excellency and Grounds of the Corpuscular Philosophy*, 1674, in M. Hunter and E. Davis (eds), *The Works of Robert Boyle*, vol. 8 (London: Pickering & Chatto, 2000), pp. 103–7, 109

[...] But when I speak of the *Corpuscular* or *Mechanical* Philosophy, I am far from meaning with the *Epicureans*, that *Atoms*, meeting together by chance in an infinite *Vacuum*, are able of themselves to produce the World, and all its Phænomena; nor with some Modern Philosophers, that, supposing God to have put into the whole Mass of Matter such an invariable quantity of Motion, he needed do no more to make the World, the material parts being able by their own unguided Motions, to cast themselves into such a System (as we call by that name): But I plead onely for such a Philosophy, as reaches but to things purely Corporeal, and distinguishing between the first *original of things*; and the subsequent *course of Nature*, teaches, concerning the *former*, not onely that God gave Motion to Matter, but that in the beginning He so guided the various Motions of the parts of it, as to contrive them into the World he design'd they should compose, (furnish'd with the *Seminal* Principles and Structures or Models of Living Creatures,) and establish'd those *Rules of Motion*, and that order amongst things Corporeal, which we are wont to call the *Laws of Nature*. And having told this as to the *former*, it may be allowed as to the *latter* to teach, That the Universe being once fram'd by God, and the Laws of Motion being setled and all upheld by His incessant concourse and general Providence; the Phænomena of the World thus constituted, are Physically produc'd by the Mechanical affections of the parts of Matter, and what they operate upon one another according to Mechanical Laws....

I. The *first* thing that I shall mention to this purpose, is the Intelligibleness or Clearness of Mechanical Principles and Explications. [...]

II. In the next place I observe, that there cannot be *fewer* Principles than the two grand ones of Mechanical Philosophy, *Matter* and *Motion*. For, Matter alone, unless it be moved, is altogether unactive; and whilst all the parts of a Body continue in one state without any Motion at all, that Body will not exercise any action, nor suffer any alteration it self, though it may perhaps modifie the action of other Bodies that move against it.

III. Nor can we conceive any Principles more *primary*, than *Matter* and *Motion*. For, either both of them were immediately created by God, or, (to add that for their sakes that would have Matter to be unproduc'd,) if *Matter* be eternal, *Motion* must either be produc'd by some Immaterial Supernatural Agent, or it must immediately flow by way of Emanation from the nature of the matter it appertains to.

IV. Neither can there be any Physical Principles more *simple* than Matter and Motion; neither of them being resoluble into any things, whereof it may be truly, or so much as tolerably, said to be compounded.

V. The next thing I shall name to recommend the Corpuscular Principle, is their great Comprehensiveness. I consider then, that the genuine and necessary effect of the sufficiently strong Motion of one part of Matter against another, is, *either* to drive it on in its intire bulk, *or* else to break or divide it into particles of determinate *Motion, Figure, Size, Posture, Rest, Order*, or *Texture.* The two first of these, for *instance*, are each of them capable of numerous varieties. For the *Figure* of a portion of matter may *either* be one of the five Regular Figures treated of by Geometricians, or some determinate *Species* of solid Figures, as that of a *Cone, Cylinder*, &c. *or* Irregular, though not perhaps Anonymous, as the Grains of Sand, Hoops, Feathers, Branches, Forks, Files, &c. And as the *Figure*, so the *Motion* of one of these particles may be exceedingly diversified, not onely by the determination to this or that part of the world, but by several other things, as particularly by the almost infinitely varying degrees of Celerity, by the manner of its progression with, or without, Rotation, and other modifying Circumstances; and more yet by the Line wherein it moves, as (besides Streight) Circular, Elliptical, Parabolical, Hyperbolical, Spiral, and I know not how many others. For, *as* later Geometricians have shewn, that those crooked Lines may be compounded of several Motions, (that is, trac'd by a Body whose motion is mixt of, and results from, two or more simpler Motions,) *so* how many more curves may, or rather may not be made by new Compositions and Decompositions of Motion, is no easie task to determine.

Now, since a *single* particle of Matter, by vertue of two onely of the Mechanical affections, that belong to it, be diversifiable so many ways; how vast a number of variations may we suppose capable of being produc'd by the Compositions and Decompositions of *Myriads* of single invisible Corpuscles, that may be contained and contex'd in one small Body, and each of them be imbued with more than two or three of the fertile Catholick Principles above mention'd? Especially since the aggregate of those Corpuscles may be farther diversifi'd by the *Texture* resulting from their Convention into a Body, which, as so made up, has its own Bigness, and Shape, and Pores, (perhaps very many, and various) and has also many capacities of acting and suffering upon the score of the place it holds among other Bodies in a World constituted as ours is: So that, when I consider the almost innumerable diversifications, that Compositions and Decompositions may make of a small number, not perhaps exceeding twenty of distinct things, I am apt to look upon those, who think the Mechanical Principles may serve indeed to give an account of the *Phænomena* of this or that particular part of Natural Philosophy, as *Staticks, Hydrostaticks*, the *Theory of the Planetary Motions*, &c. but can never be applied to all the *Phænomena* of things Corporeal; I am apt, I *say*, to look upon those, otherwise Learned, men, as I would do upon him, that should affirm, that by putting together the Letters of the *Alphabet*, one may indeed make up all the words to be found in one Book, as in *Euclid*, or *Virgil*; or in one Language, as *Latine*, or *English*; but that they can by no means suffice to supply words to all the Books of a great Library, much less to all the Languages in the world. [. . .]

And now at length I come to consider that which I observe the most to alienate other Sects from the Mechanical Philosophy; namely, that they think it pretends to have Principles so Universal and so Mathematical, that no other Physical Hypothesis can comport with it, or be tolerated by it.

But this I look upon as an easie indeed, but an important, mistake; because by this very thing, that the Mechanical Principles are so universal, and therefore applicable to so many things, they are rather fitted to *include*, than necessitated to *exclude*, any other Hypothesis that is founded in Nature, as far as it is so. And such *Hypotheses*, if prudently consider'd by a skilful and moderate person, who is rather dispos'd to unite Sects than multiply them, will be found, as far as they have Truth in them, to be either Legitimately, (though perhaps not immediately,) deducible from the Mechanical Principles, or fairly reconcilable to them. For, such Hypotheses will probably attempt to account for the *Phænomena* of Nature, *either* by the help of a determinate number of material Ingredients, such as the *Tria Prima* of the Chymists, by participation whereof other Bodies obtain their Qualities; *or* else by introducing some general Agents, as the *Platonic Soul of the World*, or the *Universal Spirit*, asserted by some Spagyrists; *or* by both these ways together.

Now to dispatch *first* those, that I named in the second place; I consider, that the chief thing, that Inquisitive Naturalists should look after in the explicating of difficult *Phænomena*, is not so much what the *Agent* is or does, as, what changes are made in the *Patient*, to bring it to exhibit the *Phænomena* that are propos'd, and by what means, and after what manner, those changes are effected. So that the *Mechanical* Philosopher being satisfied, that one part of Matter can act upon another but by vertue of Local Motion, or the effects and consequences of Local Motion, he considers, that *as*, if the propos'd Agent be not Intelligible and Physical, it can never Physically *explain* the *Phænomena; so*, if it be Intelligible and Physical, 'twill be reducible to *Matter*, and some or other of those onely Catholick affections of Matter, already often mentioned. And, the indefinite divisibility of Matter, the wonderful efficacy of Motion, and the almost infinite variety of Coalitions and Structures, that may be made of minute and insensible Corpuscles, being duly weighed, I see not why a Philosopher should think it impossible, to make out by their help the Mechanical possibility of an corporeal Agent ....

8.7 Robert Boyle, 'Dialogue on the Transmutation and Melioration of Metals' [mid- to late 1670s], in L. Principe, *The Aspiring Adept: Robert Boyle and His Alchemical Quest: Including Boyle's "Lost" Dialogue on the Transmutation of Metals* (Princeton, NJ: Princeton University Press, 1998), pp. 278–88[1]

After the whole Company had, as it were by Common Consent, continued silent for some time, which others spent in Reflections upon the Preceding

Conference, and *Pyrophilus*, in the Consideration of what he was about to Deliver; this *Virtuoso* at length stood up, and Addressing himself to the rest, 'I hope, *Gentlemen*, sayes he, that what has been already Discoursed, has Inclin'd, if not Perswaded you to Think, That the Exaltation, or Change of other *Metals* into *Gold*, is not a thing Absolutely Impossible; and, though I confess, I cannot remove all your Doubts, and Objections, or my own, by being able to Affirm to you, That I have with my own hands made Projection (as *Chymists* are wont to call the Sudden Transmutation made by a small quantity of their Admirable *Elixir*) yet I can Confirm much of what hath been Argued for the Possibility of such a sudden Change of a Metalline Body, by a Way, which, I presume, will surprize you. For, to make it credible, that other Metals are capable of being Graduated, or Exalted into Gold *by way of Projection*; I will Relate to you, that *by the like way*, Gold has been Degraded, or Imbased.'

The Novelty of this *Preamble* having much surprised the Auditory, at length, *Simplicius*, with a disdainful Smile, told *Pyrophilus*, 'That the Company would have much thanked him, if he could have assured them, That he has seen another Mettal Exalted into Gold; but, that to find a way of spoiling Gold, was not onely an Useless Discovery, but a Prejudicial Practice.'

Pyrophilus was going to make some Return to this *Animadversion*, when he was prevented by *Aristander*; who, turning himself to *Simplicius*, told him, with a Countenance and Tone that argued some displeasure; 'If *Pyrophilus* had been Discoursing to a Company of Goldsmiths, or of Merchants, your severe Reflection upon what he said would have been proper: but, you might well have forborn it, if you had considered, as I suppose he did, that he was speaking to an Assembly of *Philosophers* and *Virtuosi*, who are wont to estimate Experiments, not as they inrich Mens Purses, but their Brains, and think Knowledge especially of uncommon things very desirable, even when 'tis not accompanyed with any other thing, than the Light that still attends it, and indears it. [...]'

Pyrophilus perceiving by several signs that he needed not add any thing of Apologetical to what *Arristander* had already said for him, resumed his Discourse, by saying, 'I was going, Gentlemen, when *Simplicius* diverted me, to tell you That looking upon the Vulgar Objections that have been wont to be fram'd against the possibility of Metalline Transmutations, from the Authority and Prejudices of *Aristotle*, and the School-Philosophers, as Arguments that in such an Assembly as this need not now be solemnly discuss'd; I consider that the *difficulties* that really deserve to be call'd so, and are of weight even with Mechanical Philosophers, and Judicious Naturalists, are principally these. *First*, That the great change that must be wrought by the *Elixir*, (if there be such an Agent) is effected upon Bodies of

1 This conclusion to the 'Dialogue' was published as 'An Historical Account of the Degradation of Gold by an Anti-Elixir', 1678. Reprinted in M. Hunter and E. Davis (eds), *The Works of Robert Boyle*, vol. 9 (London: Pickering & Chatto, 2000), pp. 7–17.

so stable and almost immutable a Nature as Metals. *Next*, That this great change is said to be brought to pass in a very short time. *And thirdly*, (which is yet more strange) That this great and sudden alteration is said to be effected by a very small, and perhaps inconsiderable, proportion of transmuting Powder. To which *three* grand difficulties, I shall add *another* that to me appears, and perhaps will seem to divers of the new Philosophers, worthy to be lookt upon as a *fourth*, namely, The notable change that must by a real transmutation be made in the Specifick Gravity of the matter wrought upon: which difficulty I therefore think not unworthy to be added to the rest, because upon several tryals of my own and other men, I have found no known quality of Gold, (as its colour, malleableness, fixity, or the like) so difficult, if not so impossible, to be introduc'd into any other Metalline Matter, as the great Specifick Gravity that is peculiar to Gold. So that, Gentlemen, (concludes *Pyrophilus*) if it can be made appear that Art has produc'd an *Anti-Elixir*, (if I may so call it) or Agent that is able in a very short time, to work a very notable, though deteriorating, change upon a Metal; in proportion to which, its quantity is very inconsiderable; I see not why it should be thought impossible that Art may also make a *true Elixir*, or Powder capable of speedily Transmuting a great proportion of a baser Metal into Silver or Gold: especially if it be considered, that those that treat of these *Arcana*, confess that 'tis not every matter which may be justly called the Philosophers Stone, that is able to transmute other Metals in vast quantities; since several Writers, (and even *Lully* himself) make differing *orders* or *degrees* of the *Elixir*, and acknowledge, that a Medicine or Tincture of the first or *lowest* order will not transmute above ten times its weight of an Inferior Metal.'

[...]

The inference, saith *Pyrophilus*, I meant to make, will not detain you long; having for the main been already intimated in what you may remember I told you I design'd in the mention I was about to make of the now-recited Experiment. For without launching into difficult Speculations, or making use of disputable Hypotheses, it seems evident enough from the matter of Fact faithfully laid before you, that an Operation *very near*, if not *altogether* as strange as that which is call'd *Projection*, and in the difficultest points much of the same nature with it, may safely be admitted. For our Experiment plainly shews that Gold, though confessedly the most homogeneous, and the least mutable of Metals, may be in a very short time (perhaps not amounting to many minutes) exceedingly *chang'd*, both as to *malleableness*, *colour*, *homogeniety*, and (which is more) *specifick gravity*; and all this by so very inconsiderable a proportion of injected Powder, that since the Gold that was wrought on weighed two of our English drams, and consequently an hundred and twenty grains, an easie computation will assure us that the Medicine did thus powerfully act, according to my estimate, (which was the modestest) upon near a thousand times, (for 'twas above nine hundred and fifty times) its weight of Gold, and according to my Assistants estimate, did (as they speak) *go on* upon twelve hundred; so that if it were fit to apply to this *Anti-Elixir*, (as I formerly ventur'd to call it) what is said of the true *Elixir* by divers of the Chymical Philosophers, who will have the virtue of their Stone increas'd in such a proportion, as that at first 'twill transmute but *ten* times its weight; after the next rotation *an hundred* times, and after the next to that *a thousand times*, our Powder may in their language be stil'd *a Medicine of the third order*.

The Computation, saith *Arristander*, is very obvious, but the change of so great a proportion of Metal is so wonderful and unexampled, that I hope we shall among other things learn from it this lesson, That we ought not to be so forward as many men otherwise of great parts are wont to be, in prescribing narrow limits to the power of Nature and Art, and in condemning and deriding all those that pretend to, or believe, uncommon things in Chymistry, as either Cheats or Credulous. And therefore I hope, that though (at least in my opinion) it be very allowable to call Fables, Fables, and to detect and expose the Impostures or Deceits of ignorant or vain-glorious Pretenders to Chymical Mysteries, yet we shall not by too hasty and general censures of the sober and diligent Indigators of the *Arcana* of Chymistry, blemish (as much as in us lies) that excellent Art it self, and thereby disoblige the genuine Sons of it, and divert those that are indeed Possessors of Noble Secrets, from vouchsafing to gratifie *our* Curiosity, as we see the one of them did *Pyrophilus*'s, with the *sight* at least, of some of their highly Instructive Rarities.

I wholly approve, saith *Heliodorus* rising from his seat, the discreet and seasonable motion made by *Arristander*.

And I presume, subjoins *Pyrophilus*, that it will not be the less lik'd, if I add, That I will allow the Company to believe that as *extraordinary*, as I perceive most of you think the *Phaenomena* of the lately recited Experiment; yet I have not (because I must not do it) as yet acquainted you with the Strangest effect of our Admirable Powder.

8.8 Keynes MS 33, fols. 5*r*, *v*, Newton's supplementary remarks to the treatise 'Manna' which he received in 1675 from 'Mr. F.', in K. Figala, 'Newton as alchemist', *History of Science*, 25 (1977), pp. 134–5

'It may seem an admirable & new Paradox yt Alchemy should have concurrence wth Antiquity & Theology; ye one seeming merely humane & ye other divine; & yet Moses, yt ancient Theologus describing & expressing ye most wonderfull Architecture of this great world tells us yt ye spirit of God moved upon ye water wch was an indigested chaos, or mass created before by God wth confused earth in mixture; yet in his Alchemical extraction separation sublimation & conjunction so ordered & conjoyned again as they are manifestly seen apart & sundred.... That spiritual mover of ye first motive God, hath inspired all ye creatures of this universal world wth yt spirit of life, wch may be truly called the spirit of ye world, wch naturally moveth & secretly acteth in all creatures, giving them existence in three, viz: salt sulphur & mercury none of them being wthout their salt, the chiefest means by whose help nature bringeth forth all vegetables, minerals & animals: so yt of these three whatsoever is in nature hath its original & compo... of them so mingled wth ye four Elements that they make one body their friend. This divine Alchimy through ye operation of ye spirit (wthout wch ye Elemental & material character letter & form profiteth not) was the beginning of time... terrestrial existence by wch all things have moved and have their being, consisting of body soul & spirit whether they be vegetables minerals or Animals, only wth this difference, that ye souls of men & Angles are reasonable & immortall according to ye image of God himself, & sensual as Beasts: & such not so.

Moreover as ye omnipotent God hath in ye beginning of his divine wisdom created ye things of the heaven & of ye earth in weight number & measure, depending upon most wonderful proportions & harmony to serve ye time wch he hath appointed, so in ye fulness & last period of time wch approacheth fast as ye four Elements whereof all creatures consist, having in every of them two other Elements, the one putrifying & combustible, the other eternal & incombustible as ye Heavens... shall by Gods great Alchemy be metamorphosed & changed. For ye combustible having in them corrupt stinking feces or drossy matter wch maketh subject to corruption, & shall in ye period & general refining day be purged through fire, & then God will make new heavens & new Earth & bring all things to a crystalline clearness & will also make ye four elements perfect fixt & simple in themselves, that all things may be reduced to a Quintescence of Eternity. Thus you have a paradox & no paradox, & an Hieroglyphic plainly deciphered. For alchemy tradeth not with metalls as ignorant vulgars think, wch error hath made them distrust that noble science; but she hath also material veins of whose nature God created Handmaids to conceive & bring forth his creatures. For it is proper to God alone to create somthing of nothing, but its natures task to form that wch he hath [created].... This Philosophy both speculative & active is not only to be found in ye volume of nature but also in ye sacred scriptures...'.

8.9 Isaac Newton, *Mathematical Principles of Natural Philosophy* (The *Principia*), 1687, trans. and ed. I. B. Cohen and Anne Whitman (Berkeley, CA: University of California Press, 1999), pp. 381–3 (Author's Preface to the Reader)

SINCE THE ANCIENTS (according to Pappus) considered *mechanics* to be of the greatest importance in the investigation of nature and science and since the moderns – rejecting substantial forms and occult qualities – have undertaken to reduce the phenomena of nature to mathematical laws, it has seemed best in this treatise to concentrate on *mathematics* as it

PHILOSOPHIÆ NATURALIS PRINCIPIA MATHEMATICA.

Autore *J S. NEWTON*, *Trin. Coll. Cantab. Soc.* Mathefeos Profeffore *Lucafiano*, & Societatis Regalis Sodali.

IMPRIMATUR.
S. PEPYS, *Reg. Soc.* PRÆSES.
Julii 5. 1686.

LONDINI,

Juffu *Societatis Regiæ* ac Typis *Jofephi Streater*. Proftat apud plures Bibliopolas. *Anno* MDCLXXXVII.

relates to natural philosophy. The ancients divided *mechanics* into two parts: the *rational*, which proceeds rigorously through demonstrations, and the *practical*. *Practical mechanics* is the subject that comprises all the manual arts, from which the subject of *mechanics* as a whole has adopted its name. But since those who practice an art do not generally work with a high degree of exactness, the whole subject of *mechanics* is distinguished from *geometry* by the attribution of exactness to *geometry* and of anything less than exactness to *mechanics*. Yet the errors do not come from the art but from those who practice the art. Anyone who works with less exactness is a more imperfect mechanic, and if anyone could work with the greatest exactness, he would be the most perfect mechanic of all. For the description of straight lines and circles, which is the foundation of *geometry*, appertains to *mechanics*. *Geometry* does not teach how to describe these straight lines and circles, but postulates such a description. For *geometry* postulates that a beginner has learned to describe lines and circles exactly before he approaches the threshold of *geometry*, and then it teaches how problems are solved by these operations. To describe straight lines and to describe circles are problems, but not problems in *geometry*. *Geometry* postulates the solution of these problems from *mechanics* and teaches the use of the problems thus solved. And *geometry* can boast that with so few principles obtained from other fields, it can do so much. Therefore *geometry* is founded on mechanical practice and is nothing other than that part of *universal mechanics* which reduces the art of measuring to exact propositions and demonstrations. But since the manual arts are applied especially to making bodies move, *geometry* is commonly used in reference to magnitude, and *mechanics* in reference to motion. In this sense *rational mechanics* will be the science, expressed in exact propositions and demonstrations, of the motions that result from any forces whatever and of the forces that are required for any motions whatever. The ancients studied this part of *mechanics* in terms of the *five powers* that relate to the manual arts [i.e., the five mechanical powers] and paid hardly any attention to gravity (since it is not a manual power) except in the moving of weights by these powers. But since we are concerned with natural philosophy rather than manual arts, and are writing about natural rather than manual powers, we concentrate on aspects of gravity, levity, elastic forces, resistance of fluids, and forces of this sort, whether attractive or impulsive. And therefore our present work sets forth mathematical principles of natural philosophy. For the basic problem [*lit.* whole difficulty[1]] of philosophy seems to be to discover the forces of nature from the phenomena of motions and then to demonstrate the other phenomena from these forces. It is to these

---

1 Newton would seem to be expressing in Latin more or less the same concept that later appears in English (in query 28 of the *Opticks*) as 'the main Business of natural Philosophy'.

ends that the general propositions in books 1 and 2 are directed, while in book 3 our explanation of the system of the world illustrates these propositions. For in book 3, by means of propositions demonstrated mathematically in books 1 and 2, we derive from celestial phenomena the gravitational forces by which bodies tend toward the sun and toward the individual planets. Then the motions of the planets, the comets, the moon, and the sea are deduced from these forces by propositions that are also mathematical. If only we could derive the other phenomena of nature from mechanical principles by the same kind of reasoning! For many things lead me to have a suspicion that all phenomena may depend on certain forces by which the particles of bodies, by causes not yet known, either are impelled toward one another and cohere in regular figures, or are repelled from one another and recede. Since these forces are unknown, philosophers have hitherto made trial of nature in vain. But I hope that the principles set down here will shed some light on either this mode of philosophizing or some truer one.

In the publication of this work, Edmond Halley, a man of the greatest intelligence and of universal learning, was of tremendous assistance; not only did he correct the typographical errors and see to the making of the woodcuts, but it was he who started me off on the road to this publication. For when he had obtained my demonstration of the shape of the celestial orbits, he never stopped asking me to communicate it to the Royal Society, whose subsequent encouragement and kind patronage made me begin to think about publishing it. But after I began to work on the inequalities of the motions of the moon, and then also began to explore other aspects of the laws and measures of gravity and of other forces, the curves that must be described by bodies attracted according to any given laws, the motions of several bodies with respect to one another, the motions of bodies in resisting mediums, the forces and densities and motions of mediums, the orbits of comets, and so forth, I thought that publication should be put off to another time, so that I might investigate these other things and publish all my results together. I have grouped them together in the corollaries of prop. 66 the inquiries (which are imperfect) into lunar motions, so that I might not have to deal with these things one by one in propositions and demonstrations, using a method more prolix than the subject warrants, which would have interrupted the sequence of the remaining propositions. There are a number of things that I found afterward which I preferred to insert in less suitable places rather than to change the numbering of the propositions and the cross-references. I earnestly ask that everything be read with an open mind and that the defects in a subject so difficult may be not so much reprehended as investigated, and kindly supplemented, by new endeavors of my readers.

Trinity College, Cambridge Is. Newton
8 May 1686

OPTICKS:

OR, A

TREATISE

OF THE

Reflections, *Refractions*, *Inflections* and *Colours*

OF

LIGHT.

The FOURTH EDITION, *corrected.*

By Sir *ISAAC NEWTON*, Knt.

LONDON:

Printed for WILLIAM INNYS at the West-End of St. *Paul's*. MDCCXXX.

Title-page of the 1730 edition.

8.10 Isaac Newton, *Opticks*, 1704, 1730 edn (New York: Dover, 1952), pp. 26–33

PROP. II. THEOR. II.

The Light of the Sun consists of Rays differently Refrangible.

The PROOF by Experiments.

Exper. 3

In a very dark Chamber, at a round Hole, about one third Part of an Inch broad, made in the Shut of a Window, I placed a Glass Prism, whereby the Beam of the Sun's Light, which came in at that Hole, might be refracted upwards toward the opposite Wall of the Chamber, and there form a colour'd Image of the Sun. The Axis of the Prism (that is, the Line passing through the middle of the Prism from one end of it to the other end parallel to the edge of the Refracting Angle) was in this and the following Experiments perpendicular to the incident Rays. About this Axis I turned the Prism slowly, and saw the refracted Light on the Wall, or coloured Image of the Sun, first to descend, and then to ascend. Between the Descent and Ascent, when the Image seemed Stationary, I stopp'd the Prism, and fix'd it in that Posture, that it should be moved no more. For in that Posture the Refractions of the Light at the two Sides of the refracting Angle, that is, at the Entrance of the Rays into the Prism, and at their going out of it, were equal to one another.

So also in other Experiments, as often as I would have the Refractions on both sides the Prism to be equal to one another, I noted the Place where the Image of the Sun formed by the refracted Light stood still between its two contrary Motions, in the common Period of its Progress and Regress; and when the Image fell upon that Place, I made fast the Prism. And in this Posture, as the most convenient, it is to be understood that all the Prisms are placed in the following Experiments, unless where some other Posture is described. The Prism therefore being placed in this Posture, I let the refracted Light fall perpendicularly upon a Sheet of white Paper at the opposite Wall of the Chamber, and observed the Figure and Dimensions of the Solar Image formed on the Paper by that Light. This Image was Oblong and not Oval, but terminated with two Rectilinear and Parallel Sides, and two Semicircular Ends. On its Sides it was bounded pretty distinctly, but on its Ends very confusedly and indistinctly the Light there decaying and vanishing by degrees. The Breadth of this Image answered to the Sun's Diameter, and was about two Inches and the eighth Part of an Inch, including the Penumbra. For the Image was eighteen Feet and an half distant from the Prism, and at this distance that Breadth, if diminished by the Diameter of the Hole in the Window-shut, that is by a quarter of an Inch, subtended an Angle at the Prism of about half a Degree, which is the Sun's apparent Diameter. But the Length of the Image was about ten Inches and a quarter, and the Length of the Rectilinear Sides about eight Inches; and the refracting Angle of the Prism, whereby so great a Length was made, was 64 degrees. With a less Angle the Length of the Image was less, the Breadth remaining the same. If the Prism was turned about its Axis that way which made the Rays emerge more obliquely out of the second refracting Surface of the Prism, the Image soon became an Inch or two longer, or more; and if the Prism was turned about the contrary way, so as to make the Rays fall more obliquely on the first refracting Surface, the Image soon became an Inch or two shorter. And therefore in trying this Experiment, I was as curious as I could be in placing the Prism by the above-mention'd Rule exactly in such a Posture, that the Refractions of the Rays at their Emergence out of the Prism might be equal to that at their Incidence on it. This Prism had some Veins running along within the Glass from one end to the other, which scattered some of the Sun's Light irregularly, but had no sensible Effect in increasing the Length of the coloured Spectrum. For I tried the same Experiment with other Prisms with the same Success. And particularly with a Prism which seemed free from such Veins, and whose refracting Angle was $62\frac{1}{2}$ Degrees, I found the Length of the Image $9\frac{3}{4}$ or 10 Inches at the distance of $18\frac{1}{2}$ Feet from the Prism, the Breadth of the Hole in the Window-shut being $\frac{1}{4}$ of an Inch, as before. And because it is easy to commit a Mistake in placing the Prism in its due Posture, I repeated the Experiment four or five Times, and always found the Length of the Image that which is set down above. With another Prism of clearer Glass and better Polish, which seemed free from Veins, and whose refracting Angle was $63\frac{1}{2}$ Degrees, the Length of this Image at the same distance of $18\frac{1}{2}$ Feet was also about 10 Inches, or $10\frac{1}{8}$. Beyond these Measures for about a $\frac{1}{4}$ or $\frac{1}{3}$ of an Inch at either end of the Spectrum the Light of the Clouds seemed to be a little tinged with red and violet, but so very faintly, that I suspected that Tincture might either wholly, or in great Measure arise from some Rays of the Spectrum scattered irregularly by some Inequalities in the Substance and Polish of the Glass, and therefore I did not include it in these Measures. Now the different Magnitude of the hole in the Window-shut, and different thickness of the Prism where the Rays passed through it, and different inclinations of the Prism to the Horizon, made no sensible changes in the length of the Image. Neither did the different matter of the Prisms make any: for in a Vessel made of polished Plates of Glass cemented together in the shape of a Prism and filled with Water, there is the like Success of the Experiment according to the quantity of the Refraction. It is farther to be observed, that the Rays went on in right Lines from the Prism to the Image, and therefore at their very going out of the Prism had all that Inclination to one another from which the length of the Image proceeded, that is, the Inclination of more than two degrees and an half. And yet according to the Laws of Opticks vulgarly received, they could not possibly be so much inclined to one another.* For let EG [*Fig.* 13] represent the Window-shut, F the hole made

* See our Author's *Lectiones Opticæ*, Part. I. Sect. 1. §5.

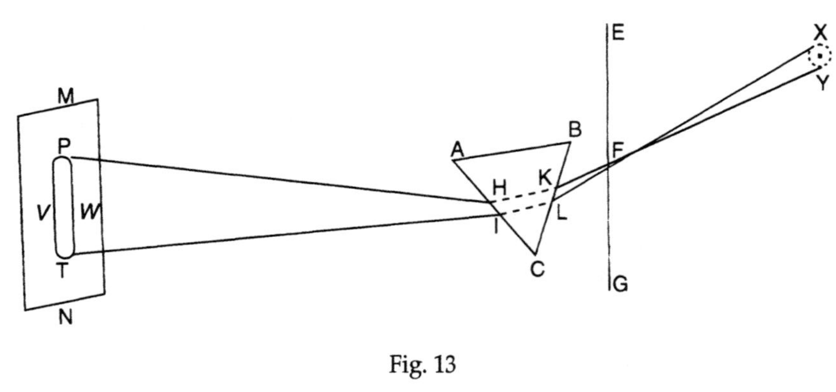

Fig. 13

therein through which a beam of the Sun's Light was transmitted into the darkened Chamber, and ABC a Triangular Imaginary Plane whereby the Prism is feigned to be cut transversely through the middle of the Light. Or if you please, let ABC represent the Prism it self, looking directly towards the Spectator's Eye with its nearer end: And let XY be the Sun, MN the Paper upon which the Solar Image or Spectrum is cast, and PT the Image it self whose sides towards *v* and *w* are Rectilinear and Parallel, and ends towards P and T Semicircular. YKHP and XLIT are two Rays, the first of which comes from the lower part of the Sun to the higher part of the Image, and is refracted in the Prism at K and H, and the latter comes from the higher part of the Sun to the lower part of the Image, and is refracted at L and I. Since the Refractions on both sides of the Prism are equal to one another, that is, the Refraction at K equal to the Refraction at I, and the Refraction at L equal to the Refraction at H, so that the Refractions of the incident Rays at K and L taken together, are equal to the Refractions of the emergent Rays at H and I taken together: it follows by adding equal things to equal things, that the Refractions at K and H taken together, are equal to the Refractions at I and L taken together, and therefore the two Rays being equally refracted, have the same Inclination to one another after Refraction which they had before; that is, the Inclination of half a Degree answering to the Sun's Diameter. For so great was the inclination of the Rays to one another before Refraction. So then, the length of the Image PT would by the Rules of Vulgar Opticks subtend an Angle of half a Degree at the Prism, and by Consequence be equal to the breadth *vw*; and therefore the Image would be round. Thus it would be were the two Rays XLIT and YKHP, and all the rest which form the Image P*w*T*v*, alike refrangible. And therefore seeing by Experience it is found that the Image is not round, but about five times longer than broad, the Rays which going to the upper end P of the

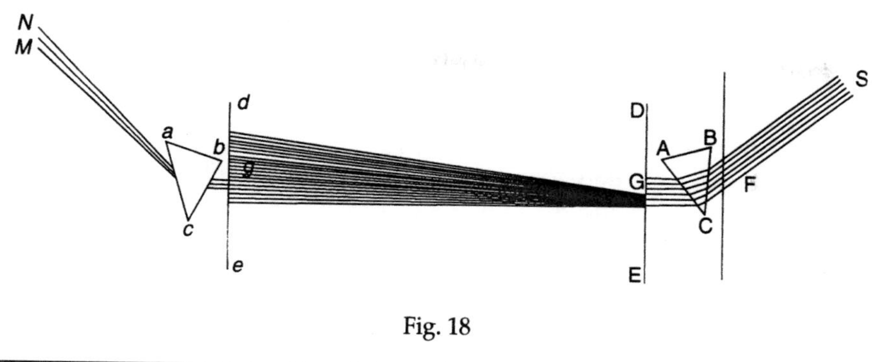

Fig. 18

Image suffer the greatest Refraction, must be more refrangible than those which go to the lower end T, unless the Inequality of Refraction be casual.

This Image or Spectrum PT was coloured, being red at its least refracted end T, and violet at its most refracted end P, and yellow green and blue in the intermediate Spaces. Which agrees with the first Proposition, that Lights which differ in Colour, do also differ in Refrangibility. The length of the Image in the foregoing Experiments, I measured from the faintest and outmost red at one end, to the faintest and outmost blue at the other end, excepting only a little Penumbra, whose breadth scarce exceeded a quarter of an Inch, as was said above. [. . .]

Exper. 6

In the middle of two thin Boards I made round holes a third part of an Inch in diameter, and in the Window-shut a much broader hole being made to let into my darkened Chamber a large Beam of the Sun's Light; I placed a Prism behind the Shut in that beam to refract it towards the opposite Wall, and close behind the Prism I fixed one of the Boards, in such manner that the middle of the refracted Light might pass through the hole made in it, and the rest be intercepted by the Board. Then at the distance of about twelve Feet from the first Board I fixed the other Board in such manner that the middle of the refracted Light which came through the hole in the first Board, and fell upon the opposite Wall, might pass through the hole in this other Board, and the rest being intercepted by the Board might paint upon it the coloured Spectrum of the Sun. And close behind this Board I fixed another Prism to refract the Light which came through the hole.

Illustration.

Let F [in *Fig.* 18] be the wide hole in the Window-shut, through which the Sun shines upon the first Prism ABC, and let the refracted Light fall upon the middle of the Board DE, and the middle part of that Light upon the hole G made in the middle part of that Board. Let this trajected part of that Light fall again upon the middle of the second Board *de*, and there paint such an oblong coloured Image of the Sun as was described in the third Experiment. By turning the Prism ABC slowly to and fro about its Axis, this Image will be made to move up and down the Board *de*, and by this means all its parts from one end to the other may be made to pass successively through the hole *g* which is made in the middle of that Board. In the mean while another Prism *abc* is to be fixed next after that hole *g*, to refract the trajected Light a second time. And these things being thus ordered, I marked the places M and N of the opposite Wall upon which the refracted Light fell, and found that whilst the two Boards and second Prism remained unmoved, those places by turning the first Prism about its Axis were changed perpetually. For when the lower part of the Light which fell upon the second Board *de* was cast through the hole *g*, it went to a lower place M on the Wall and when the higher part of that Light was cast through the same hole *g*, it went to a higher place N on the Wall, and when any intermediate part of the Light was cast through that hole, it went to some place on the Wall between M and N. The unchanged Position of the holes in the Boards, made the Incidence of the Rays upon the second Prism to be the same in all cases. And yet in that common Incidence some of the Rays were more refracted, and others less. And those were more refracted in this Prism, which by a greater Refraction in the first Prism were more turned out of the way, and therefore for their Constancy of being more refracted are deservedly called more refrangible.

Chapter Nine
Scientific Academies across Europe

9.1 Introduction to the *Saggi* of the Accademia del Cimento, 1667, in W. E. K. Middleton, *The Experimenters: A Study of the Accademia del Cimento* (Baltimore, MD: John Hopkins University Press, 1971), pp. 87, 89–92

TO THE MOST SERENE FERDINAND II, GRAND DUKE OF TUSCANY

MOST SERENE LORD:

The printing of the first samples of the experiments in natural philosophy that have been made for many years in our Academy with the assistance and under the continued protection of the Most Serene Prince Leopold Your Highness' brother, will itself carry to those regions of the world in which virtue shines most brightly, new evidence of the great munificence of Your Highness and call back towards you with a new sense of gratitude the true lovers of the fine arts and the most noble sciences. We ourselves are the more greatly moved to a more devoted acknowledgment, for we have been so much nearer, enjoying the strong influence of your beneficent hand. Meanwhile, with the favor of your protection, with the stimulus of your ability and your own taste and inclination, and above all with the honor of your presence, sometimes coming into the Academy and sometimes calling it into your Royal apartments, you have given the Academy its reputation and its fervor and at the same time enlarged the progress of our studies.

These considerations very easily teach us our clear duty of consecrating to Your Highness' lofty name this first fruit of our labors, inasmuch as nothing can come from us in which Your Highness might have a greater part and which it would thus be more proper to offer you or which would come nearer to meriting the good fortune of your generous approval. Truly, for the superabundance of so many and so remarkable favors we feel no greater desire than to see ourselves thus strictly bound to Your Highness; not because we willingly carry the burden of such precious and valued obligations, but because we should wish only to be able to offer you something that is not your own, so that we might at least flatter ourselves that we had brought you some slight recompense and expressed some thanks of our own choice to Your Highness that would not be entirely your own or from necessity. But now we are perforce content to have in our hearts such just and proper sentiments, since the fruit of these new philosophical speculations is so strongly rooted in Your Highness' protection that not only what our Academy produces today but everything that matures in the most famous schools of Europe, and that will come to light in succeeding ages, will likewise properly be due to Your Highness as the gift of your beneficence. For as long as sun, stars, and planets shine and the heavens endure, there will remain the glorious memory of one who contributed so much with the power of his most happy favors to such new and stupendous discoveries and to the opening of an untrodden path to the more exact investigation of truth. Although we have so little to offer, there is something that sharpens our respectful gratitude. This indeed is the pleasure with which we bear our poverty, while everything overflows in the greater abundance of glory for Your Highness, who, having yourself done whatever new, good, and great things will ever be found in the wealth of the sciences, have enfeebled in others all power to equal you. So much and no more are we in a position to offer Your Highness. Filled with reverence and homage and begging your continued protection, we pray God to grant you the highest prosperity and greatness.

Your Highness' very humble servants,
THE ACADEMICIANS OF THE CIMENTO

Florence, 14 July 1667
The Secretary for the *Saggi*

[. . .]

Preface

Without any doubt the first born of all the creatures of the Divine wisdom was the idea of Truth, to which the eternal Architect held so closely in his plan for the building of the universe that nothing whatever was made with even the slightest alloy of falsehood. But when later on, in the contemplation of such a high and perfect structure, man acquired too immoderate a desire to understand the marvellous power of it and to take the measure and proportions of such a beautiful harmony, then, wishing to enter too completely into the truth, he came to create an indefinite number of falsehoods. This had no other cause but the desire for those wings that Nature did not want to give him, perhaps fearing that she might at some time be surprised in the preparation of her most stupendous works. On such wings he began to soar, though oppressed by the weight of his material body, gaining the strength to rise higher than the gamut of perceptible things, and so tried to fix himself in a light such as is no longer light when it enters the eyes, but, fading, darkens and changes color.

This was how the first seeds of false opinions arose from human rashness. It is not that the brightness of God's beautiful creations is at all obscured by this, or that they are in any way spoilt by their commerce with it; for these defects all lie in human ignorance, where they have their root. Meanwhile, man, improperly fitting causes to effects, takes their true essence neither from one nor the other, but putting them together forms a false science within his own mind. But it is not that the sovereign beneficence of God, when creating our souls, does not let them glance for an instant, so to speak, at the immense treasure of His eternal wisdom, adorning them with the first gleams of truth as with precious gems. We see that this is true from the notions, preserved in our spirits, that we cannot have learned here, but must inevitably have gathered from another place.

But it is indeed our misfortune that these most noble jewels, badly held in the setting of a soul still too tender, just as it comes down into its earthly house and wraps itself in this common clay, at once fall out of their fastenings and are so befouled as to be worthless, until at length by assiduous and eager study they are brought back to their places. Now this is exactly what the soul is trying to do in the investigation of natural things. [...] Now here, where we are no longer permitted to step forward, there is nothing better to turn to than our faith in experiment. As one may take a heap of loose and unset jewels and seek to put them back one after another into their setting, so experiment, fitting effects to causes and causes to effects – though it may not succeed at the first throw, like Geometry – performs enough so that by trial and error it sometimes succeeds in hitting the target.

But we must proceed with great caution lest too much faith in experiment should deceive us, since before it shows us manifest truth, after lifting the first veils of more evident falsehood, it always makes visible certain misleading appearances that seem to be true. These are those indistinct outlines that show through the last veils that cover the lovely image of Truth more closely, through the fineness of which she sometimes appears so vividly pictured that anyone might say that all was discovered. Here, then, it is our business to have a masterly understanding of the ways of truth and falsehood and to use our own judgment as shrewdly as we can, so as to see clearly whether it is or is not. There is no doubt that to be able to do better we must at some time have seen Truth unveiled, an advantage possessed only by those who have acquired some taste for the study of geometry.

Yet besides trying new experiments, it is not less useful to search among those already made, in case any might be found that might in any way have counterfeited the pure face of Truth. So it has been the aim of our Academy, besides doing whatever experiments occurred to us, to make them, either from fruitful curiosity or for comparison, on those things that have been done or written about by others, as unfortunately we see errors being established by things that go by the name of experiments. This was precisely what first moved the most perspicaceous and indefatigable mind of His Highness Prince Leopold of Tuscany, who, as a recreation after the diligent management of affairs and the ceaseless anxieties that arise from his exalted state, began to exercise his intellect upon the steep road of the most noble kinds of knowledge. Therefore it was quite easy for His Highness' sublime understanding to see how the trust in great authors is most often inimical to talented men. Either from excessive trust or from reverence for a name, such people dare not consider doubtful what is conjectured on such great authority. His Highness decided that it should be the task of his great spirit to verify the value of their assertions by wiser and more exact experiments: and when the proof or the refutation had been obtained, to present it as a gift, so precious and so desirable, to whomsoever is anxious to discover Truth.

These prudent precepts of His Highness our protector, embraced by the Academy with the veneration and esteem that they deserved, have not had it as their aim to make them indiscreet censors of other laborious scholars, or presumptuous dispensers of truth or error; but have mainly been intended to encourage other people to repeat the same experiments with the greatest rigor, just as we have sometimes dared to do to those of others, even though in publishing these first examples we have for the most part abstained from this, the better to confirm, with proper regards to all our readers, the sincerity of our dispassionate and respectful opinions. In fact, to give full scope to such a noble and useful enterprise we should wish for nothing else but a free communication from the various Societies, scattered as they are today throughout the most illustrious and notable region of Europe. These, opening such a profitable mutual correspondence with the same purpose of reaching such important ends, would all go on searching with the same liberty, according to their means, and participating in Truth. For our part, we shall contribute to this work with the greatest sincerity and ingenuousness.

Wherever we have reported the experiments of others, we have always cited the authors, as far as they were known to us, and have often freely confessed being much helped in these trials, which we did not after all succeed in bringing to as happy an end. But as the clearest proof of the open sincerity of our procedure, all may see the liberality with which we have always shared them with anyone passing through these parts who showed a desire to enjoy some account of them, whether as an act of courtesy, or because he esteemed learning, or from the incentive of noble curiosity. This we have done since the first days of our Academy, founded in the year 1657, during which early days were discovered, if not all, then the greater part of the things of which these examples are now being printed.

If, after all, someone perceives that among the things that we give out as ours some are to be found that were first imagined and published by others, this will never be our fault; for as we can neither know all nor see all, nobody should marvel at the correspondence between our intellects and those of others, just as we really are not surprised at the correspondence between theirs and ours.

We should certainly not wish anyone to persuade himself that we have the presumption to bring out a finished work, or even a perfect pattern of a great experimental history, knowing well that more time and labor are going to be required for such a purpose. Everyone may perceive this from the very title we have given this book, merely of examples. We should never even have published this without the powerful stimulus that we have had from honorable people whose kind entreaties persuaded us to suffer the shame of printing such imperfect beginnings.

Finally, before everything else we protest that we would never wish to pick a quarrel with anyone, entering into subtle disputes or vain contradictions; and if sometimes in passing from one experiment to another, or for any other reason whatever, some slight hint of speculation is given, this is always to be taken as the opinion or private sentiment of the academicians, never that of the Academy, whose only task is to make experiments and to tell about them. For such was our first intention and also the purpose of that exalted Person who with his particular protection and great knowledge led us along the way, and whose wise and prudent counsel we have always exactly and regularly obeyed.

9.2 Thomas Sprat, *History of the Royal Society*, 1667: (a) 'On the History of the Royal Society: A Model of Their Whole Design', (b) 'On the History of the Royal Society: Their Course of Inquiry' (London: Routledge, Kegan and Paul, 1966), (a) pp. 61–76, (b) pp. 76–100

(a) 'On the History of the Royal Society: A Model of Their Whole Design'

Their purpose is, in short, to make faithful *Records*, of all the Works of *Nature*, or *Art*, which can come within their reach: that so the present Age, and posterity, may be able to put a mark on the Errors, which have been strengthened by long prescription: to restore the Truths, that have lain neglected: to push on those, which are already known, to more various uses: and to make the way more passable, to what remains unreveal'd. This is the compass of their Design. And to accomplish this, they have indeavor'd, to separate the knowledge of *Nature*, from the colours of *Rhetorick*, the devices of *Fancy*, or the delightful deceit of *Fables*. They have labor'd to inlarge it, from being confin'd to the custody of a few, or from servitude to private interests. They have striven to preserve it from being over-press'd by a confus'd heap of vain, and useless particulars; or from being straitned and bounded too much up by General Doctrines. They have try'd, to put it into a condition of perpetual increasing; by settling an inviolable correspondence between the hand, and the brain. They have studi'd, to make it, not onely an Enterprise of one season, or of some lucky opportunity; but a business of time; a steddy, a lasting, a popular, an uninterrupted Work. They have attempted, to free it from the Artifice, and Humors, and Passions of Sects; to render it an Instrument, whereby Mankind may obtain a Dominion over *Things*, and not onely over one anothers *Judgements*. And lastly, they have begun to establish these Reformations in Philosophy, not so much, by any solemnity of Laws, or ostentation of Ceremonies; as by solid Practice, and examples: not, by a glorious pomp of Words; but by the silent, effectual, and unanswerable Arguments of real Productions. [...]

As for what belongs to the *Members* themselves, that are to constitute the *Society*: It is to be noted, that they have freely admitted Men of different Religions, Countries, and Professions of Life. [...]

That the *Church of England* ought not to be apprehensive, of this free converse of various Judgments, I shall afterwards manifest at large. For the present, I shall franckly assert; that our *Doctrine*, and *Discipline*, will be so far from receiving damage by it; that it were the best way to make them universally embrac'd, if they were oftner brought to be canvas'd amidst all sorts of dissenters. [...]

By their *naturalizing* Men of all Countries, they have laid the beginnings of many great advantages for the future. For by this means, they will be able, to settle a *constant Intelligence*, throughout all civil Nations; and make the *Royal Society* the general *Banck*, and Free-port of the World: A policy, which whether it would hold good, in the *Trade of England*, I know not: but sure it will in the *Philosophy*. We are to overcome the mysteries of all the Works of Nature; and not onely to prosecute such as are confin'd to one Kingdom, or beat upon one shore. [...]

By their admission of Men of all *professions*, these *two* Benefits arise: The one, that every Art, and every way of life already establish'd, may be secure of receiving no damage by their Counsels. A thing which all new Inventions ought carefully to consult. It is in vain, to declare against the profit of the most, in any change that we would make. [...] But the other benefit is, that by this equal Balance of all Professions, there will no one particular of them overweigh the other, or make the *Oracle* onely speak their *private* fence: which else it were impossible to avoid. [...]

But, though the *Society* entertains very many men of *particular Professions*; yet the farr greater Number are *Gentlemen*, free, and unconfin'd. By the help of this, there was hopefull Provision made against *two corruptions* of Learning, which have been long complain'd of, but never remov'd: The *one*, that *Knowledge* still degenerates, to consult *present profit* too soon; the *other*, that *Philosophers* have bin always *Masters*, & *Scholars*; some imposing, & all the other submitting; and not as equal observers without dependence. [...]

The second Error, wich is hereby endeavour'd to be remedied, is, that the Seats of Knowledg, have been for the most part heretofore, not *Laboratories*, as they ought to be; but onely *Scholes*, where some have *taught*, and all the rest subscrib'd. The consequences of this are very mischievous. [...]

The very inequality of the Titles of *Teachers*, and *Scholars*, does very much suppress, and tame mens Spirits; which though it should be proper for Discipline and Education; yet is by no means consistent with a free Philosophical Consultation. [...]

These therefore are the *Qualities*, which they have principally requir'd, in those, whom they admitted: still reserving to themselves a power of *increasing*, or keeping to their number, as they saw occasion. By this means, they have given assurance of an eternal quietness, and moderation, in their experimental progress; because they allow themselves to differ in the weightiest matter, even in the *way of Salvation* itself. By this they have taken care, that nothing shall be so remote, as to escape their reach: because some of their *Members* are still scattered abroad, in most of the habitable parts of the Earth. By this, they have provided, that no profitable thing shall seem too mean for their consideration, seeing they have some amongst them, whose life is employ'd about *little* things, as well as *great*. By this they have broken down the partition wall, and made a fair entrance, for *all conditions of men* to engage in these Studies; which were heretofore affrighted from them, by a groundless apprehension of their chargeableness, and difficulty. Thus they have form'd that *Society*, which intends a *Philosophy*, for the use of *Cities*, and not for the retirements of *Schools*, to resemble the *Cities* themselves: which are compounded of all sorts of men, of the *Gown*, of the *Sword*, of the *Shop*, of the *Field*, of the *Court*, of the *Sea*; all mutually assisting each other.

(b) 'On the History of the Royal Society: Their Course of Inquiry'

Let us next consider what *course of Inquiry* they take, to make all their Labours unite for the service of man-kind: And here I shall insist on their *Expence*, their *Instruments*, their *Matter*, and their *Method*.

Of the Stock, upon which their Expence has been hitherto defraid, I can say nothing, that is very *magnificent*: seeing they have rely'd upon no more than some small *Admission-money*, and *weekly Contributions* amongst themselves. Such a *Revenue* as this, can make no great sound, nor amount to any *vast summ*. But yet, I shall say this for it, that it was the onely way, which could have been begun, with a security of success, in that condition of things. The *publick Faith* of *Experimental Philosophy*, was not then strong enough, to move Men and Women of all conditions, to bring in their Bracelets and Jewels, towards the carrying of it on. Such affections as those may be rais'd by a mis-guided zeal; but seldom, or never, by calm and unpassionate Reason. It was therefore well ordain'd, that the first Benevolence should come from the *Experimenters themselves*. If they had speedily at first call'd for *mighty Treasures*; and said aloud, that their Enterprise requir'd the *Exchequer of a Kingdom*; they would onely have been contemn'd, as vain *Projectors*. So ready is man-kind, to suspect all new undertakings to be Cheats, and *Chimæraes*; especialy, when they seem *chargeable*: [...]

The suspicion, which is so natural to mens breasts, [...] could not any way harm the *Royal Societies* establishment: seeing its first claims, and pretensions were so modest. And yet I shall presume to assure the World; that what they shall raise on these mean Foundations, will be more answerable to the largeness of their intentions, than to the *narrowness* of their beginnings. [...] To evidence this; I think it may be calculated, that since the *Kings* Return, there have been more *Acts of Parliament*, for the *clearing* and *beautifying* of Streets, for the *repayring* of *Highwayes*, for the *cutting* of *Rivers*, for the *increase* of *Manufactures*, for the setting on foot the Trade of Fishing, and many other such Publick Works, to adorn the State; than in divers Ages before. This *General Temper* being well weigh'd; it cannot be imagin'd, that the *Nation* will withdraw its assistance from the *Royal Society* alone; which does not intend to stop at some *particular benefit*, but goes to the root of *all noble Inventions*, and proposes an infallible course to make *England* the glory of the Western world. [...]

In their *Method of Inquiring*, I will observe, how they have behav'd themselves, in things that might be brought within their *own Touch and Sight*: and how in those, which are so remote, and hard to be come by, that about them, they were forc'd to trust *the reports of others*.

In the first kind: I shall lay it down, as their *Fundamental Law*, that whenever they could possibly get to *handle* the subject, the *Experiment* was still perform'd by some of the *Members* themselves. [...] And the Task was divided amongst them, by one of these two ways. First, it was sometimes referr'd to some *particular men*, to make choice of what *Subject* they pleased, and to follow their own humour in the *Trial*; the *expence* being still allow'd from the general Stock. [...]

Or else secondly, the *Society* it self made the distribution, and deputed whom it thought fit for the prosecution of such, or such Experiments. And this they did, either by allotting the *same Work* to *several* men, separated one from another; or else by *joyning* them into *Committees*. [...] By this *union* of *eyes*, and *hands* there do these *advantages* arise. Thereby there will be a full *comprehension* of the object in *all* its appearances; and so there will be a mutual communication of the light of one *Science* to another: whereas *single labours* can be but as a prospect taken upon one side. And also by this fixing of several mens thoughts upon one thing, there will be an excellent cure for that *defect*, which is almost unavoidable in great *Inventors*. It is the custom of such earnest, and powerful minds, to do wonderful things in the *beginning*; but shortly after, to be overborn by the multitude, and weight of their own thoughts; then to yield, and cool by little and little; and at last grow weary, and even to loath that, upon which they were at first the most eager. [...]

For this, the best provision must be, to join many men together; for it cannot be imagin'd, that they should be all so violent, and fiery: and so by this mingling of Tempers, the *Impetuous* men, not having the whole burthen on them, may have leisure for intervals to recruit their first heat; and the more *judicious*, who are not so soon possess'd with such raptures, may carry on the others strong conceptions, by soberer degrees, to a full accomplishment.

This they have practis'd in such things, whereof the matter is common; and wherein they may repeat their labours as they please. But in *forein*, and *remote* affairs, their *Intentions*, and their *Advantages* do farr exceed all others. For these, they have begun to settle a *correspondence* through all Countreys; and have taken such order, that in short time, there will scarce a Ship come up the *Thames*, that does not make some return of *Experiments*, as well as of *Merchandize*.

This their care of an *Universal Intelligence*, is befriended by *Nature* its self, in the situation of *England*: For, lying so, as it does, in the passage between the Northern parts of the World, and the *Southern*; its *Ports* being open to all Coasts, and its *Ships* spreading their Sails in all Seas; it is thereby necessarily made, not onely *Mistress* of the *Ocean*, but the most proper *Seat*, for the advancement of *Knowledg*. From the *positions* of Countreys, arise not only their several shapes, manners, customs, colours, but also their *different Arts*, and *Studies*. The *Inland* and *Continent*, we see do give Laws, to Discourse, to Habits, to Behaviour: but those that border upon the *Seas*, are most properly seated, to bring home matter for *new Sciences*, and to make the same proportion of Discoveries above others, in the *Intellectual* Globe, as they have done in the *Material*. [. . .]

Their *Matter*, being thus collected, has been brought before their *weekly meetings*, to undergo a just and a full examination. In them their principal endeavours have been, that they might enjoy the benefits of a *mix'd Assembly*, which are largeness of Observation, and diversity of Judgments, without the mischiefs that usually accompany it, such as confusion, unsteddiness, and the little animosities of divided Parties. That they have avoided these dangers for the time past; there can be no better proof, than their constant practice; wherein they have perpetually preserv'd a singular sobriety of debating, slowness of consenting, and moderation of dissenting. Nor have they been onely free from *Faction*, but from the very *Causes*, and *beginnings* of it. It was in vain for any man amongst them to strive to preferr himself before another; or to seek for any great glory from the subtilty of his Wit; seeing it was the inartificial process of the *Experiment*, and not the *Acuteness* of any Commentary upon it, which they have had in veneration. There was no room left, for any to attempt, to heat their own, or others minds, beyond a due temper; where they were not allow'd to expatiate, or amplifie, or connect specious arguments together. [. . .]

Towards the first of these ends, it has been their usual course, when they themselves appointed the *Trial*, to propose one week, some particular *Experiments*, to be prosecuted the next; and to debate beforehand, concerning all things that might conduce to the better carrying them on. In this *Præliminary Collection*, it has been the custom, for any of the *Society*, to urge what came into their thoughts, or memories concerning them; either from the observations of others, or from *Books*, or from their own *Experience*, or even from common *Fame* it self. And in performing this, they did not exercise any great rigour of choosing, and distinguishing between *Truths* and *Falshoods*: but a mass altogether as they came; the certain Works, the Opinions, the Ghesses, the Inventions, with their different Degrees and Accidents, the Probabilities, and Problems, the general Conceptions, the miraculous Stories, the ordinary Productions, the changes incident to the same Matter in several places, the Hindrances, the Benefits, of *Airs*, or *Seasons*, or *Instruments*; and whatever they found to have been begun, to have fail'd, to have succeeded, in the Matter which was then under their Disquisition.

This is a most necessary preparation, to any that resolve to make a perfect search. For they cannot but go blindly, and lamely, and confusedly about the business, unless they have first laid before them a full *Account* of it. [...]

Those, to whom the conduct of the *Experiment* is committed, being dismiss'd with these advantages, do (as it were) carry the eyes, and the imaginations of the whole company into the *Laboratory* with them. And after they have perform'd the *Trial*, they bring all the *History* of its *process* back again to the *test*. Then comes in the second great Work of the *Assembly*; which is to *judg*, and *resolve* upon the matter of *Fact*. In this part of their imployment, they us'd to take an exact view of the repetition of the whole course of the *Experiment*; here they observ'd all the *chances*, and the *Regularities* of the proceeding; what *Nature* does willingly, what constrain'd; what with its own power, what by the succours of Art; what in a constant rode, and what with some kind of sport and extravagance; industriously marking all the various shapes into which it turns it self, when it is persued, and by how many secret passages it at last obtains its end; never giving it over till the whole *Company* has been fully satisfi'd of the certainty and constancy; or, on the otherside, of the absolute impossibility of the effect. This *critical*, and *reiterated scrutiny* of those things, which are the plain objects of their eyes; must needs put out of all reasonable dispute, the reality of those operations, which the *Society* shall positively determine to have succeeded. If any shall still think it a just *Philosophical liberty*, to be jealous of resting on their credit: they are in the right; and their *dissentings* will be most thankfully receiv'd, if they be establish'd on solid works, and not onely on *prejudices*, or *suspicions*. To the *Royal Society* it will be at any time almost as acceptable, to be *confuted*, as to *discover*: seeing, by this means, they will accomplish their main *Design*: others will be inflam'd: many more will labour; and so the *Truth* will be obtain'd between them: which may be as much promoted by the *contentions* of hands, and eyes; as it is commonly injur'd by those of Tongues. However, that men may not hence undervalue their *authority*, because they themselves are not willing to impose, and to usurp a *dominion* over their reason; I will tell them, that there is not any one thing, which is now approv'd and practis'd in the World, that is confirm'd by stronger evidence, than this, which the *Society* requires; except onely the *Holy Mysteries of our Religion*. In almost all other matters of *Belief*, of *Opinion*, or of *Science*; the assurance, whereby men are guided, is nothing near so firm as this.

9.3 Letter from Henry Oldenburg to A. Auzout, 24 May 1666, in A. Rupert Hall and M. B. Hall (eds and trans.), *The Correspondence of Henry Oldenburg*, 13 vols, 1965–86 (Madison: University of Wisconsin Press, 1966), vol. 3, pp. 141–2

To Mr. Auzout, 24 May 1666

Sir,

As my esteem for your merits did not allow me to propose you for election into our Society[1] in an other than numerous meeting, when the opportunity to do so presented itself recently I did not fail to seize it, with a measure of success entirely proportionate to your merits and my expectations. This illustrious assembly yesterday chose you as a member with an applause that was as general as your knowledge and honesty (known to all who know you) deserved. By this you see, Sir, that this company holds the opinion that political and national differences ought not to impede the philosophical exchange nor shut the door upon the appreciation of the sciences and of virtue. Moreover you will judge that these gentlemen are convinced that you will not fail to use your splendid talents to contribute a very considerable amount to the progress of their design, which is none other than the search for truth and the welfare of humanity. They themselves hope that the French people, so rich in great men and fine intellects, will at last establish a similar society under the auspices and beneficence of so great a king as Louis XIV; for however great, wise, and magnificent, he cannot count himself a complete Prince until he has founded a truly philosophical Academy, that is, one which will study Nature rather than Aristotle and will strive to join with others the power of its intelligence and industry in order to restore men to their primitive condition through a sound knowledge of the works of the Almighty Creator, that is to say, to restore to men that splendid dominance over created beings from which they fell.

I am constantly inclined to believe that something of this kind will be accomplished in your kingdom and even during the reign of the present king, when I think of his wisdom, generosity, and power.

If my conjectures prove to be correct (as I wish with all my heart) I have no doubt that all other nations, however little civilized, will be diligent in following so good an example and in making a philosophical alliance with France and England in order to achieve by their joint efforts the objects of a design as noble and useful as that is. That, Sir, is the conviction of

Your very humble and affectionate
servant
H. O.

---

1 Oldenburg proposed the French physicist A. Auzout (1622–91).

Chapter Ten
The Reception of Newtonianism across Europe

10.1 Richard Bentley, *A Confutation of Atheism from the Origin and Frame of the World*, 1693, in *Isaac Newton's Papers and Letters on Natural Philosophy and Related Documents*, ed. I. B. Cohen, 2nd edn (Cambridge, MA: Harvard University Press, 1978), pp. 315–22

When we first enter'd upon this Topic, the demonstration of God's Existence from the Origin and Frame of the World, we offer'd to prove four Propositions.

1. That this present System of Heaven and Earth cannot possibly have subsisted from all Eternity.

2. That Matter consider'd generally, and abstractly from any particular Form and Concretion, cannot possibly have been eternal: Or, if Matter could be so; yet Motion cannot have coexisted with it eternally, as an inherent property and essential attribute of Matter. These two we have already established in the preceding Discourse; we shall now shew in the third place,

3. That, though we should allow the Atheists, that Matter and Motion may have been from everlasting; yet if (as they now suppose) there were once no Sun nor Starrs nor Earth nor Planets; but the Particles, that now constitute them, were diffused in the mundane Space in manner of a Chaos without any concretion and coalition; those dispersed Particles could never of themselves by any kind of Natural motion, whether call'd Fortuitous or Mechanical, have conven'd into this present or any other like Frame of Heaven and Earth.

A

Confutation of Atheism

FROM THE

Origin and Frame of the WORLD.

The Third and Last PART.

A

SERMON

Preached at

St Mary-le-Bow,

DECEMBER the 5th. 1692.

Being the *Eighth* of the Lecture Founded by the Honourable *ROBERT BOYLE*, Esquire.

By *RICHARD BENTLEY*, M. A.

Chaplain to the Right Reverend Father in God, *EDWARD*, Lord Bishop of *Worcester*.

LONDON,

Printed for *H. Mortlock* at the *Phœnix* in St. *Paul's* Church-yard. 1693.

I. And first as to that ordinary Cant of illiterate and puny Atheists, the *fortuitous or casual concourse of Atoms*, that compendious and easy Dispatch of the most important and difficult affair, the Formation of a World; (besides that in our next undertaking it will be refuted all along) I shall now briefly dispatch it, from what hath

Serm. V. p. 6, 7. been formerly said concerning the true notions of Fortune and Chance. Whereby it is evident, that in the Atheistical Hypothesis of the World's production, Fortuitous and Mechanical must be the self-same thing. Because *Fortune* is no real entity nor physical essence, but a mere relative signification, denoting only this; That such a thing said to fall out by Fortune, was really effected by material and necessary Causes; but the Person, with regard to whom it is called Fortuitous, was ignorant of those Causes or their tendencies, and did not design nor foresee such an effect. This is the only allowable and genuine notion of the word Fortune. But thus to affirm, that the World was made *fortuitously*, is as much as to say, That before the World was made, there was some Intelligent Agent or Spectator; who designing to do something else, or expecting that something else would be done with the Materials of the World, there were some occult and unknown motions and tendencies in Matter, which mechanically formed the World beside his design or expectation. Now the Atheists, we may presume, will be loth to assert a fortuitous Formation in this proper sense and meaning; whereby they will make Understanding to be older than Heaven and Earth. Or if they should so assert it; yet, unless they will affirm that the Intelligent Agent did dispose and direct the inanimate Matter, (which is what we would bring them to) they must still leave their Atoms to their mechanical Affections; not able to make one step toward the production of a World beyond the necessary Laws of Motion. It is plain then, that *Fortune*, as to the matter before us, is but a synonymous word with Nature and Necessity. It remains that we examin the adequate meaning of *Chance*; which properly signifies,

Serm. V. p. 12, 13. That all events called Casual, among inanimate Bodies, are mechanically and naturally produced according to the determinate figures and textures and motions of those Bodies; with this negation only, That those inanimate Bodies are not conscious of their own operations, nor contrive and cast about how to bring such events to pass. So that thus to say, that the World was made *casually* by the concourse of Atoms, is no more than to affirm, that the Atoms composed the World mechanically and fatally; only they were not sensible of it, nor studied and consider'd about so noble an undertaking. For if Atoms formed the World according to the essential properties of Bulk, Figure and Motion, they formed it *mechanically*; and if they formed it mechanically without perception and design, they formed it *casually*. So that this negation of Consciousness being all that the notion of Chance can add to that of Mechanism; We, that do not dispute this matter with the Atheists, nor believe that Atoms ever acted by Counsel and Thought, may have leave to consider the several names of *Fortune* and *Chance* and *Nature* and *Mechanism*, as one and the same Hypothesis. Wherefore once for all to overthrow all possible Explications which Atheists have or may assign for the formation of the World, we will undertake to evince this following Proposition:

II. That the Atoms or Particles which now constitute Heaven and Earth, being once separate and diffused in the Mundane Space, like the supposed *Chaos*, could never *without a God by their Mechanical affections* have convened into this present Frame of Things or any other like it.

Which that we may perform with the greater clearness and conviction; it will be necessary, in a discourse about the Formation of the World, to give you a brief account of some of the most principal and systematical *Phænomena*, that occurr in the World now that it is formed.

(1.) The most considerable *Phænomenon* belonging to Terrestrial Bodies is the general action of *Gravitation*, whereby All known Bodies in the vicinity of the Earth do tend and press toward its Center; not only such as are sensibly and evidently Heavy, but even those that are comparatively the Lightest, and even in their proper place, and natural Elements, (as they usually speak) as Air gravitates even in Air and Water in Water. This hath been demonstrated and experimentally proved beyond contradiction, by several ingenious Persons of the present Age, but by none so perspicuously and copiously and accurately, as by the Honourable Founder of this Lecture in his incomparable Treatises of the *Air* and *Hydrostaticks*.

Mr. *Boyle's* Physicom. Exp. of Air. Hydrostat. Paradoxes.

(2.) Now this is the constant Property of *Gravitation*; That the weight of all Bodies around the Earth is ever proportional to the Quantity of their Matter: As for instance, a Pound weight (examin'd Hydrostatically) of all kinds of Bodies, though of the most different forms and textures, doth always contain an equal quantity of solid Mass or corporeal Substance. This is the ancient Doctrine of the *Epicurean* Physiology, then and since very probably indeed, but yet precariously asserted: But it is lately demonstrated and put beyond controversy by that very excellent and divine Theorist Mr. *Isaac Newton*, to whose most admirable sagacity and industry we shall frequently be obliged in this and the following Discourse.

Lucret. lib. I.

Newton Philos. Natur. Princ. Math. lib. 3. prop. 6.

I will not entertain this Auditory with an account of the Demonstration; but referring the Curious to the Book itself for full satisfaction, I shall now proceed and build upon it as Truth solidly established, *That all Bodies weigh according to their Matter*; provided only that the compared Bodies be at equal distances from the Center toward which they weigh. Because the further they are removed from the Center, the lighter they are: decreasing gradually and uniformly in weight, in a duplicate proportion to the Increase of the Distance.

(3.) Now since Gravity is found proportional to the Quantity of Matter, there is a manifest Necessity of admitting a *Vacuum*, another principal Doctrine of the *Atomical* Philosophy. Because if there were every-where an absolute plenitude and density without any empty pores and interstices between the Particles of Bodies, then all Bodies of equal dimensions would contain an equal Quantity of Matter; and consequently, as we have shewed before, would be equally ponderous: so that Gold, Copper, Stone, Wood, *&c.* would have all the same specifick weight; which Experience assures us they have not: neither would any of them descend in the Air, as we all see they do; because, if all Space was Full, even the Air would be as dense and specifically as heavy as they. If it be said, that, though the difference of specifick Gravity may proceed from variety of Texture, the lighter Bodies being of a more loose and porous composition, and the heavier more dense and compact; yet an æthereal subtile Matter, which is in a perpetual motion, may penetrate and pervade the minutest and inmost Cavities of the closest Bodies, and adapting it self to the figure of every Pore, may adequately fill them; and so prevent all Vacuity, without increasing the weight: To this we answer; That that subtile Matter it self must be of the same Substance and Nature with all other Matter, and therefore It also must weigh proportionally to its Bulk; and as much of it as at any time is comprehended within the Pores of particular Body must gravitate jointly with that Body: so that if the Presence of this æthereal Matter made an absolute Fullness, all Bodies of equal dimensions would be equally heavy: which being refuted by experience, it necessarily follows, that there is a Vacuity; and that (notwithstanding some little objections full of cavil and sophistry) mere and simple Extension or Space hath a quite different nature and notion from the real Body and impenetrable Substance.

10.2 Letter from I. Newton to R. Bentley, 11 February 1693, in *Isaac Newton's Papers and Letters on Natural Philosophy and Related Documents*, ed. I. B. Cohen, 2nd edn (Cambridge, MA: Harvard University Press, 1978), pp. 310–12

To Mr. BENTLEY, *at the Palace*
at Worcester

SIR,

THE Hypothesis of deriving the Frame of the World by mechanical Principles from Matter evenly spread through the Heavens, being inconsistent with my System, I had considered it very little before your Letters put me upon it, and therefore trouble you with a Line or two more about it, if this comes not too late for your Use.

In my former I represented that the diurnal Rotations of the Planets could not be derived from Gravity, but required a divine Arm to impress them. And tho' Gravity might give the Planets a Motion of Descent towards the Sun, either directly or with some little Obliquity, yet the transverse Motions by which they revolve in their several Orbs, required the divine Arm to impress them according to the Tangents of their Orbs. I would now add, that the Hypothesis of Matter's being at first evenly spread through the Heavens, is, in my Opinion, inconsistent with the Hypothesis of innate Gravity, without a supernatural Power to reconcile them, and therefore it infers a Deity. For if there be innate Gravity, it is impossible now for the Matter of the Earth and all the Planets and Stars to fly up from them, and become evenly spread throughout all the Heavens, without a supernatural Power; and certainly that which can never be hereafter without a supernatural Power, could never be heretofore without the same Power.

You queried, whether Matter evenly spread throughout a finite Space, of some other Figure than spherical, would not in falling down towards a central Body, cause that Body to be of the same Figure with the whole Space, and I answered, yes. But in my Answer it is to be supposed that the Matter descends directly downwards to that Body, and that that Body has no diurnal Rotation.

This, Sir, is all I would add to my former Letters.

I am,
Your most humble
Servant,

Cambridge,
Feb. 11, 1693.

IS. NEWTON.

10.3 John Ray, *The Wisdom of God Manifested in the Works of The Creation*, 1691, in D. C. Goodman (ed.), *Science and Religious Belief, 1600–1900: A Selection of Primary Sources* (Bristol: John Wright and Open University Press, 1973), pp. 210–19

The Providence of Nature is wonderful in a *Camel*, or *Dromadary*, both in the Structure of his Body, and in the Provision that is made for the Sustenance of it. Concerning the first, I shall instance only in the Make of his Foot, the Sole whereof, as the *Parisian Academists* do observe, is flat and broad, being very fleshy, and covered only with a thick, soft, and somewhat callous Skin, but very fit and proper to travel in sandy Places; such as are the Desarts of *Africk* and *Asia*. We thought (say they) that this Skin was like a living Sole, which wore not with the Swiftness and the Continuance of the March, for which this *Animal* is almost indefatigable. And it may be this Softness of the Foot, which yields and fits itself to the Ruggedness and Uneavenness of the Roads, does render the Feet less capable of being worn, than if they were more solid.

As to the Second, the Provision that is made for their Sustenance in their continued Travels over sandy Desarts, the same Academists observe, That at the Top of the second Ventricle (for they are ruminant Creatures, and have four Stomachs) there were several square Holes, which were the Orifices of about twenty Cavities, made like Sacks placed between the two Membranes, which do compose the Substance of this Ventricle. The View of these Sacks made us to think, that they might well be the Reservatories, where *Pliny* says, That Camels do a long time keep the Water, which they drink in great Abundance when they meet with it, to supply the Wants which they may have thereof in the dry Desarts, wherein they are used to travel; and where it is said, that those that do guide them, are sometimes forced, by Extremity of Thirst, to open their Bellies, in which they do find Water.

THE
WISDOM of GOD
Manifested in the
WORKS
OF THE
CREATION.

In TWO PARTS.

VIZ.
The Heavenly Bodies, Elements, Meteors, Fossils, Vegetables, Animals, (Beasts, Birds, Fishes, and Insects); more particularly in the Body of the Earth, its Figure, Motion, and Consistency; and in the admirable Structure of the Bodies of Man, and other Animals; as also in their Generation, *&c.* With Answers to some Objections.

By *JOHN RAY*, late Fellow of the *Royal Society*.

The NINTH EDITION, Corrected.

LONDON:
Printed by WILLIAM and JOHN INNYS, Printers to the *Royal Society*, at the West-End of St. *Paul's*.
M DCC XXVII.

That such an Animal as this, so patient of long Thirst, should be bred in such droughty and parched Countries, where it is of such eminent Use for travelling over those dry and sandy Desarts, where no Water is to be had sometimes in two or three Days Journey, no candid and considerable Person but must needs acknowledge to be an Effect of Providence and Design.

I should now proceed to answer some Objections which might be made against the Wisdom and Goodness of God in the Contrivance and Governance of the World, and all Creatures therein contained....

Object. *A wise Agent acts for Ends. Now what End can there be of creating such a vast Multitude of Insects, as the World is filled with; most of which seems to be useless, and some also noxious and pernicious to Man, and other Creatures?*

Answ. To this I shall Answer; 1. As to the Multitude of *Species*, or Kinds. 2. As to the Number of *Individuals*, in each Kind.

First, As to the Multitude of *Species*, (which we must needs acknowledge to be exceeding great, they being not fewer, perchance more than Twenty Thousand.) I answer there were so many made.

1. To manifest and display the Riches of the Power and Wisdom of God, *Psalm* civ. 24. *The Earth is full of thy Riches; so is this great and wide Sea, wherein are Things creeping innumerable*, &c. We should be apt to think too meanly of those Attributes of our Creator, should we be able to come to an End of all his Works, even in this Sublunary World. And therefore, I believe, never any Man yet did, never any Man shall, so long as the World endures, by his utmost Industry, attain to the Knowledge of all the *Species* of Nature. Hitherto we have been so far from it, that in Vegetables, the Number of those which have been discovered this last Age, hath far exceeded that of all those which were known before. ... The World is so richly furnished and provided, that Man need not fear Want of Employment, should he live to the Age of *Methuselah*, or ten times as long....

2. Another Reason why so many Kinds of Creatures were made, might be to exercise the contemplative Faculty of Man; which is in nothing so much pleas'd, as in Variety of Objects. We soon grow weary of one Study; and if all the Objects of the World could be comprehended by us, we should, with *Alexander*, think the World too little for us, and grow weary of running in a Round of seeing the same Things. New Objects afford us great Delight, especially if found out by our own Industry. ... Thus God is pleased, by reserving Things to be found out by our Pains and Industry, to provide us Employment most delightful and agreable to our Natures and Inclinations.

3. Many of these Creatures may be useful to us, whose Uses are not yet discovered, but reserved for the Generations to come, as the Uses of some we now know are but of late Invention, and were unknown to our Forefathers. And this must needs be so, because, as I said before, the World is too great for any Man, or Generation of Men, by his, or their utmost Endeavours, to discover and find out all its Store and Furniture, all its Riches and Treasures.

Secondly, As to the Multitude of Individuals in each Kind of *Insect*. I Answer,

1. It is designed to secure the Continuance and Perpetuity of the several Species; which, if they did not multiply exceedingly, scarce any of them could escape the Ravine of so many Enemies as continually assault and prey upon them, but would endanger to be quite destroyed and lost out of the World.

2. This vast Multitude of Insects is useful to Mankind, if not immediately, yet mediately. It cannot be denied, that Birds are of great Use to us; their Flesh affording us a good Part of our Food, and that the most delicate too, and their other Parts Physick, not excepting their very Excrements. Their Feathers serve to stuff our Beds and Pillows, yielding us soft and warm Lodging, which is no small Convenience and Comfort to us, especially in these Northern Parts of the World. Some of them have also been always employed by Military Men in Plumes, to adorn their Crests, and render them formidable to their Enemies. Their Wings and Quills are made use of for Writing-Pens, and to brush and cleanse our Rooms, and their Furniture. Besides, by their melodious Accents they gratify our Ears; by their beautiful Shapes and Colours, they delight our Eyes, being very ornamental to the World, and rendring the Country where the Hedges and Woods are full of them, very pleasant and chearly, which without them wou'd be no less lonely and melancholy. Not to mention the Exercise, Diversion, and Recreation, which some of them give us.

Now Insects supply Land-Birds the chiefest Part of their Sustenance: Some, as the entire *Genus* of *Swallows*, live wholly upon them, as I could easily make out, did any Man deny or doubt of it: And not *Swallows* alone, but also *Wood-peckers*, if not wholly, yet chiefly; and all other Sorts of Birds partly, especially in Winter-time, when Insects are their main Support, as appears by dissecting their Stomachs.

As for young Birds, which are brought up in the Nest by the old, they are fed chiefly, if not solely, by Insects. And therefore for the Time when Birds for the most Part breed in the Spring, when there are Multitudes of Caterpillars to be found on all Trees and Hedges. Moreover, it is very remarkable, that of many such Birds, as when grown up, feed almost wholly upon Grain, the young ones are nourish'd by Insects. For Example, *Pheasants* and *Partridges*, which are well known to be granivorous Birds, the Young live only, or mostly, upon Ants Eggs. Now Birds, being of a hot Nature, are very voracious Creatures, and eat abundantly, and therefore there had need be an infinite Number of Insects produced for their Sustenance. Neither do Birds alone, but many Sorts of Fishes, feed upon Insects, as is well known to Anglers, who bait their Hooks with them. Nay, which is more strange, divers Quadrupeds feed upon Insects, and some live wholly upon them, as two Sorts of *Tamunduus* upon Ants, which therefore are called in *English* Ant-Bears; the *Camelion* upon Flies; the Mole upon Earth-Worms: The *Badger* also lives chiefly upon Beetles, Worms, and other Insects.

Here we may take Notice by the way, That because so many Creatures live upon Ants and their Eggs, Providence hath so order'd it, that they should be the most numerous of any Tribe of Insects that we know. [. . .]

As for noxious Insects, why there should be so many of them produced, if it be demanded,

I answer, 1. That many that are noxious to us, are salutary to other Creatures; and some that are Poison to us, are Food to them. So we see the

Poultry-kind feed upon *Spiders*: Nay, there is scarce any noxious Insect, but one Bird or other eats it, either for Food or Physick. For many, nay, most of those Creatures, whose Bite, or Sting, is poisonous, may safely be taken entire into the Stomach. And therefore it is no wonder, that not only the *Ibis* of *Egypt*, but even *Storks* and *Peacocks*, prey upon and destroy all Sorts of *Serpents*, as well as *Locusts* and *Caterpillars*.

2. Some of the most venomous and pernicious of Insects afford us noble Medicines, as *Scorpions*, *Spiders*, and *Cantharides*.

3. These Insects seldom make use of their offensive Weapons, unless assaulted or provoked in their own Defence, or to revenge an Injury. Let them but alone, and annoy them not, nor disturb their Young, and, unless accidentally, you shall seldom suffer by them.

Lastly, God is pleased sometimes to make use of them as Scourges, to chastize or punish wicked Persons, or Nations, as he did *Herod* and the *Egyptians*. No Creature so mean and contemptible, but God can, when he pleases, produce such Armies of them, as no humane Force is able to conquer, or destroy; but they shall of a sudden consume and devour up all the Fruits of the Earth, and whatever might serve for the Sustenance of Man, as *Locusts* have often been observed to do.

Did these Creatures serve for no other Use, as they do many; yet those that make them an Objection against the Wisdom of God, may... as well upbraid the Prudence and Policy of a State for keeping Forces, which generally are made up of very rude and insolent People, which yet are necessary, either to suppress Rebellions, or punish Rebels, and other disorderly and vicious Persons, and keep the World in quiet.

10.4 Roger Cotes, Preface to the 2nd edn of the *Principia*, 1713, in *Isaac Newton: The 'Principia': Mathematical Principles of Natural Philosophy*, ed. I. B. Cohen and A. Whitman (Berkeley, CA: University of California Press, 1999), pp. 391–3, 397–9

[...] We have at last reached the point where it must be acknowledged that the earth and the sun and all the celestial bodies that accompany the sun attract one another. Therefore every least particle of each of them will have its own attractive force in proportion to the quantity of matter, as was shown above for terrestrial bodies. And at different distances their forces will also be in the squared ratio of the distances inversely; for it is mathematically demonstrated that particles attracting by this law must constitute globes attracting by the same law.

The preceding conclusions are based upon an axiom which is accepted by every philosopher, namely, that effects of the same kind – that is, effects whose known properties are the same – have the same causes, and their properties which are not yet known are also the same. For if gravity is the cause of the fall of a stone in Europe, who can doubt that in America the cause of the fall is the same? If gravity is mutual between a stone and the earth in Europe, who will deny that it is mutual in America? If in Europe the attractive force of the stone and the earth is compounded of the attractive forces of the parts, who will deny that in America the force is similarly compounded? If in Europe the attraction of the earth is propagated to all kinds of bodies and to all distances, why should we not say that in America it is propagated in the same way? All philosophy is based on this rule, inasmuch as, if it is taken away, there is then nothing we can affirm about things universally. The constitution of individual things can be found by observations and experiments; and proceeding from there, it is only by this rule that we make judgments about the nature of things universally.

Now, since all terrestrial and celestial bodies on which we can make experiments or observations are heavy, it must be acknowledged without exception that gravity belongs to all bodies universally. And just as we must not conceive of bodies that are not extended, mobile, and impenetrable, so we should not conceive of any that are not heavy. The extension, mobility, and impenetrability of bodies are known only through experiments; it is in exactly the same way that the gravity of bodies is known. All bodies for which we have observations are extended and mobile and impenetrable; and from this we conclude that all bodies universally are extended and mobile and impenetrable, even those for which we do not have observations. Thus all bodies for which we have observations are heavy; and from this we conclude that all bodies universally are heavy, even those for which we do not have observations. If anyone were to say that the bodies of the fixed stars are not heavy, since their gravity has not yet been observed, then by the same argument one would be able to say that they are neither extended nor mobile nor impenetrable, since these properties of the fixed stars have not yet been observed. Need I go on? Among the primary qualities of all bodies universally, either gravity will have a place, or extension, mobility, and impenetrability will not. And the nature of things either will be correctly explained by the gravity of bodies or will not be correctly explained by the extension, mobility, and impenetrability of bodies.

I can hear some people disagreeing with this conclusion and muttering something or other about occult qualities. They are always prattling on and on to the effect that gravity is something occult, and that occult causes are to be banished completely from philosophy. But it is easy to answer them: occult causes are not those causes whose existence is very clearly demonstrated by observations, but only those whose existence is occult, imagined, and not yet proved. Therefore gravity is not an occult cause of celestial motions, since it has been shown from phenomena that this force really exists. Rather, occult causes are the refuge of those who assign the governing of these motions to some sort of vortices of a certain matter utterly fictitious and completely imperceptible to the senses.

But will gravity be called an occult cause and be cast out of natural philosophy on the grounds that the cause of gravity itself is occult and not yet found? Let those who so believe take care lest they believe in an absurdity that, in the end, may overthrow the foundations of all philosophy. For causes generally proceed in a continuous chain from compound to more simple; when you reach the simplest cause, you will not be able to proceed any further. Therefore no mechanical explanation can be given for the simplest cause; for if it could, the cause would not yet be the simplest. Will you accordingly call these simplest causes occult, and banish them? But at the same time the causes most immediately depending on them, and the causes that in turn depend on these causes, will also be banished, until philosophy is emptied and thoroughly purged of all causes.

Some say that gravity is preternatural and call it a perpetual miracle. Therefore they hold that it should be rejected, since preternatural causes have no place in physics. It is hardly worth spending time on demolishing this utterly absurd objection, which of itself undermines all of philosophy. For either they will say that gravity is not a property of all bodies – which cannot be maintained – or they will assert that gravity is preternatural on the grounds that it does not arise from other affections of bodies and thus not from mechanical causes. Certainly there are primary affections of bodies, and since they are primary, they do not depend on others. Therefore let them consider whether or not all these are equally preternatural, and so equally to be rejected, and let them consider what philosophy will then be like.

There are some who do not like all this celestial physics just because it seems to be in conflict with the doctrines of Descartes and seems scarcely capable of being reconciled with these doctrines. They are free to enjoy their own opinion, but they ought to act fairly and not deny to others the same liberty that they demand for themselves. Therefore, we should be allowed to adhere to the Newtonian philosophy, which we consider truer, and to prefer causes proved by phenomena to causes imagined and not yet proved. It is the province of true philosophy to derive the natures of things from causes that truly exist, and to seek those laws by which the supreme artificer willed to establish this most beautiful order of the world, not those laws by which he could have, had it so pleased him. For it is in accord with reason that the same effect can arise from several causes somewhat different from one another; but the true cause will be the one from which the effect truly and actually does arise, while the rest have no place in true philosophy. [...]

...Those who hold that the heavens are filled with fluid matter, but suppose this matter to have no inertia, are saying there is no vacuum but in fact are assuming there is one. For, since there is no way to distinguish a fluid matter of this sort from empty space, the whole argument comes down to the names of things and not their natures. But if anyone is so devoted to matter that he will in no way admit a space void of bodies, let us see where this will ultimately lead him.

For such people will say that this constitution of the universe as everywhere full, which is how they imagine it, has arisen from the will of God, so that a very subtle aether pervading and filling all things would be there to facilitate the operations of nature; this cannot be maintained, however, since it has already been shown from the phenomena of comets that this aether has no efficacy. Or they will say that this constitution has arisen from the will of God for some unknown purpose, which ought not to be said either, since a different constitution of the universe could equally well be established by the same argument. Or finally they will say that it has not arisen from the will of God but from some necessity of nature. And so at last they must sink to the lowest depths of degradation, where they have the fantasy that all things are governed by fate and not by providence, that matter has existed always and everywhere of its own necessity and is infinite and eternal. On this supposition, matter will also be uniform everywhere, for variety of forms is entirely inconsistent with necessity. Matter will also be without motion; for if by necessity matter moves in some definite direction with some definite velocity, by a like necessity it will move in a different direction with a different velocity; but it cannot move in different directions with different velocities; therefore it must be without motion. Surely, this world – so beautifully diversified in its forms and motions – could not have arisen except from the perfectly free will of God, who provides and governs all things.

From this source, then, have all the laws that are called laws of nature come, in which many traces of the highest wisdom and counsel certainly appear, but no traces of necessity. Accordingly we should not seek these laws by using untrustworthy conjectures, but learn them by observing and experimenting. He who is confident that he can truly find the principles of physics, and the laws of things, by relying only on the force of his mind and the internal light of his reason should maintain either that the world has existed from necessity and follows the said laws from the same necessity, or that although the order of nature was constituted by the will of God, nevertheless a creature as small and insignificant as he is has a clear understanding of the way things should be. All sound and true philosophy is based on phenomena, which may lead us – however unwilling and reluctant – to principles in which the best counsel and highest dominion of an all-wise and all-powerful being are most clearly discerned; these principles will not be rejected because certain men may perhaps not like them. These men may call the things that they dislike either miracles or occult qualities, but names maliciously given are not to be blamed on the things themselves, unless these men are willing to confess at last that philosophy should be based on atheism. Philosophy must not be overthrown for their sake, since the order of things refuses to be changed. [...]

Therefore Newton's excellent treatise will stand as a mighty fortress against the attacks of atheists; nowhere else will you find more effective ammunition against that impious crowd. This was understood long ago, and was first splendidly demonstrated in learned discourses in English and in Latin, by a man of universal learning and at the same time an outstanding patron of the arts, Richard Bentley, a great ornament of his time and of our academy, the worthy and upright master of our Trinity College. I must confess that I am indebted to him on many grounds; you as well, kind reader, will not deny him due thanks. For, as a long-time intimate friend of our renowned author (he considers being celebrated by posterity for this friendship to be of no less value than becoming famous for his own writings, which are the delight of the learned world), he worked simultaneously for the public recognition of his friend and for the advancement of the sciences. Therefore, since the available copies of the first edition were extremely rare and very expensive, he tried with persistent demands to persuade Newton (who is distinguished as much by modesty as by the highest learning) and finally – almost scolding him – prevailed upon Newton to allow him to get out this new edition, under his auspices and at his own expense, perfected throughout and also enriched with significant additions. He authorized me to undertake the not unpleasant duty of seeing to it that all this was done as correctly as possible.

Cambridge, 12 May 1713

Roger Cotes,
Fellow of Trinity College,
Plumian Professor of Astronomy
and Experimental Philosophy

10.5 Extracts from letters relating to *The Leibniz–Clarke correspondence*, 1715–16, ed. and introd. H. G. Alexander (Manchester University Press, 1956), pp. 184–8: (a) Leibniz to Conti, November–December 1715, (b) Newton to Conti, 26 February 1716, (c) Leibniz to Conti, 9 April 1716, (d) Leibniz to Wolff, 23 December 1715

The following extracts contain the passages of philosophical interest in the letters which Leibniz and Newton wrote to Conti [and Wolff] about each other at the same time as the Leibniz–Clarke exchange was taking place.

(a) Leibniz to Conti, November–December 1715

His [i.e. Newton's] philosophy appears to me rather strange and I cannot believe it can be justified. If every body is heavy it follows necessarily (whatever his supporters may say and however passionately they deny it) that gravity will be a scholastic occult quality or else the effect of a miracle. I did at one time convince M. Bayle that whatever cannot be explained by the nature of created things is miraculous. It is not sufficient to say: God has made such a law of Nature, therefore the thing is natural. It is necessary that the law should be capable of being fulfilled by the nature of created things. If, for example, God were to give to a free body the law of revolving around a certain centre, he would either have to join to it other bodies which by their impulsion made it always stay in its circular orbit, or to put an angel at its heels; or else he would have to concur extraordinarily in its motion. For naturally it would go off along the tangent. God acts continually on his creatures in the conservation of their natures, and this conservation is a continual production of that which is in itself perfection. He is *intelligentia supramundana* because he is not the soul of the world and has no need of a sensorium.

I do not find the existence of a vacuum proved by the argument of M. Newton or his followers, any more than the universal gravity which they suppose, or the existence of atoms. One can only accept the existence of a vacuum or of atoms, if one has very limited views. M. Clarke contests the opinion of the Cartesians who believe that God cannot destroy one part of matter to make a vacuum, but I am astonished that he does not see that if space is a substance different from God, the same difficulty occurs. Now to say that God is space, is to give him parts. Space is the order of co-existents and time is the order of successive existents. They are things true but ideal, like numbers.

Matter itself is not a substance but only *substantiatum*, a well-founded phenomenon, and which does not mislead one at all if one proceeds by reasoning according to the ideal laws of arithmetic, geometry, dynamics, & etc. Everything I put forward in that seems to me proved. As to dynamics or the doctrine of forces, I am astonished that M. Newton and his followers believe that God has made his machine so badly that unless he affects it by some extraordinary means, the watch will very soon cease to go. This is to have very narrow ideas of the wisdom and power of God. I call extraordinary every operation of God demanding something other than the conservation of the natures of created things. I believe the metaphysics of these gentlemen *a narrow one* and their mathematics *arrivable* enough; this does not prevent me from estimating very highly the physico-mathematical meditations of M. Newton; and you would, Sir, do a great service to the public, if you would persuade this able man to give us his recent conjectures in physics. I strongly approve of his method of drawing from phenomena what can be drawn without making any suppositions, even if sometimes this is only drawing conjectural consequences. However, when the *data* are not sufficient, it is permissible (as one does sometimes in deciphering) to imagine hypotheses, and if they are good ones to hold them provisionally, waiting for new experiments to bring us *nova data* and for what Bacon calls *experimenta crucis*, in order to choose between hypotheses. As I have learned that certain Englishmen have misrepresented my philosophy in their *Transactions*, I have no doubt that, with what I send you here, I can be justified. I am strongly in favour of the experimental philosophy but M. Newton is departing very far from it when he claims that all matter is heavy (or that every part of matter attracts every other part) which is certainly not proved by experiments, as M. Huygens has already asserted; gravitating matter could not itself have that weight of which it is the cause and M. Newton adduces no experiment or sufficient reason for the existence of a vacuum or of atoms or for the general mutual attraction. And because we do not yet know perfectly and in detail how gravity is produced or elastic force or magnetic force, this does not give us any right to make of them scholastic occult qualities or miracles; but it gives us still less right to put bounds to the wisdom and power of God and to attribute to him a sensorium and such things. Furthermore I am astonished that the followers of M. Newton have produced nothing to show that their master has taught them a good method; I have been more fortunate in my disciples.

(b) Newton to Conti, 26 February 1716

Hitherto Mr. Leibnitz avoided returning an answer to the *Commercium Epistolicum*[1] by pretending that he had not seen it. And now he avoids it by telling you that the English shall not have the pleasure to see him return an answer to their slender reasonings (as he calls them) and by endeavouring to engage me in disputes about philosophy and about solving of problems;[2] both of which are nothing to the question.

As for philosophy, he colludes in the signification of words, calling those things miracles which create no wonder; and those things occult qualities, whose causes are occult, though the qualities themselves be manifest; and those things the souls of men, which do not animate their bodies. His *harmonia praestabilita* is miraculous and contradicts the daily experience of all mankind; every man finding in himself a power of seeing with his eyes, and moving his body by his will. He prefers hypotheses to arguments of induction drawn from experiments, accuses me of opinions which are not mine; and instead of proposing questions to be examined by experiments before they are admitted into philosophy, he proposes hypotheses to be admitted and believed before they are examined. But all this is nothing to the *Commercium Epistolicum*.

(c) Leibniz to Conti, 9 April 1716

I do not wish here to go into detail about what M. Newton says rather bitterly against my philosophy and in favour of his own. This is not the place. I call a *miracle* any event which can only occur through the power of the Creator, its reason not lying in the nature of created things; and when nevertheless one would attribute it to the qualities or powers of created

---

1 The *Commercium Epistolicum* was a collection of letters and manuscripts relating to the Newton–Leibniz calculus controversy, selected and published together with a report by a special committee of the Royal Society.

2 Leibniz had ended his letter to Conti by propounding a problem whose solution involved the calculus 'in order to feel the pulse of the English analysts'.

things, then I call this quality a *scholastic occult quality*; that is one that it is impossible to render clear, such as a primitive heaviness; for the occult qualities which are not chimerical are those whose cause we do not know but do not exclude. And I call the soul of man that simple substance which perceives what happens in the human body and whose desires or acts of will are followed by the exertions of the body. I do not prefer hypotheses to arguments drawn by induction from experiments; but things are sometimes passed off as general inductions when they only consist in particular observations, and sometimes passed off as a hypothesis when they are capable of demonstration. The idea which M. Newton gives here of my pre-established harmony is not that which many able men outside England have, and I cannot believe that you yourself have had a similar idea or have it now unless you have just changed your views.

(d) Leibniz to Wolff, 23 December 1715

Her Royal Highness the Princess of Wales, who read my *Theodicy* very attentively and was delighted with it, recently, as she herself told me, argued about it with a certain English clergyman who has access to court. She attacked Newton and his followers for holding that God needs to correct and reanimate his machine. She thinks that my opinion according to which everything proceeds well from what is pre-established, and there is no need of divine correction but only divine sustentation, is more conformable with the perfections of God. He gave her Royal Highness a paper written in English in which he tries to defend Newton's view and attack mine. He would impute to me the denial of divine governance, if everything proceeds well by itself. But he does not realize that the divine governance of natural things consists in sustentation, and must not be taken in an anthropological sense. I replied immediately and sent my reply to the Princess.

10.6 W. J. 'sGravesande, *Mathematical Elements of Natural Philosophy*, 3rd edn, London, 1726, pp. xii–xviii (Preface to vol. 1)

The Study of Natural Philosophy is not however to be contemned, as built upon an unknown Foundation. The Sphere of humane Knowledge is bounded within a narrow Compass; and he, that denies his Assent to every Thing but Evidence, wavers in Doubt every Minute; and looks upon many Things as unknown which the Generality of People never so much as call in Question. But rightly to distinguish Things known, from Things unknown, is a Perfection above the Level of human Mind. Though many Things in Nature are hidden from us; yet what is set down in Physics, as a Science, is undoubted. From a few general Principles numberless particular Phænomena or Effects are explained, and deduced by Mathematical Demonstration. For, the comparing of Motion, or in other Words, of Quantities, is the continual Theme; and whoever will go about that Work any other Way, than by Mathematical Demonstrations, will be sure to fall into Uncertainties at least, if not in Errors.

How much soever then may be unknown in Natural Philosophy, it still remains a vast, certain and very useful Science. It corrects an infinite Number of Prejudices concerning natural Things, and divine Wisdom; and, as we examine the Works of GOD continually, sets that Wisdom before our Eyes; and there is a wide Difference, betwixt knowing the divine Power and Wisdom by a Metaphysical Argument, and beholding them with our Eyes every Minute in their Effects. It appears then sufficiently, what is the End of Physics, from what Laws of Nature the Phænomena are to be deduced, and wherefore, when we are once come to the general Laws, we cannot penetrate any further into the Knowledge of Causes. There remains only to discourse of the Method of searching after those Laws; and to prove that the three *Newtonian* Laws delivered in the first Chapter of this Work ought to be followed.

The first is, *That we ought not to admit any more Causes of Natural Things, than what are true, and sufficient to explain their Phænomena*. The first Part of this Rule plainly follows from what has been said above. The other cannot be called in Question by any that owns the Wisdom of the Creator. If one Cause suffices, it is needless to superadd another; especially, if it be considered, that an Effect from a double Cause is never exactly the same with an Effect from a single one. Therefore we are not to multiply Causes, till it appears one single Cause will not do the Business.

In order to prove the following Rules, we must premise some general Reflections.

We have already said, that Mathematical Demonstrations have no Standard to be judged by, but their Conformity with our Ideas; and when the Question is about Natural Things, the first Requisite is, that our Ideas agree with those Things, which cannot be proved by any Mathematical Demonstration. And yet as we have Occasion to reason of Things themselves every Moment, and of those Things nothing can be present to our Minds besides our Ideas, upon which our Reasonings immediately turn; it follows, that GOD has established some Rules, by which we may judge of the Agreement of our Ideas with the Things themselves.

All Mathematical Reasonings turn upon the Comparison of Quantities, and their Truth is evidenc'd by implying a Contradiction in a contrary Proposition. A rectilineal Triangle, for Instance, whose three Angles are not equal to two right ones, is a Thing impossible. When the Question is not about the Comparison of Quantities, a contrary Proposition is not always impossible. It is certain, for Instance, that *Peter* is living, though it is as certain that he might have died Yesterday. Now there being numberless Cases of that Kind, where one may affirm or deny with equal Certainty; there follows, that there are many Reasonings very certain, tho' altogether different from the Mathematical Ones. And they evidently follow from the Establishment of Things, and therefore from the predetermined Will of

GOD. For by forcing Men upon the Necessity of pronouncing concerning the Truth or Falsehood of a Proposition; he plainly shews they must assent to Agreements, which their Judgments necessarily acquiesce in; and whoever reasons otherwise, does not think worthily of GOD.

To return to *Physics*; we are in this Science to judge by our Senses, of the Agreement that there is betwixt Things and our Ideas. The Extension and Solidity of Matter, for Instance, asserted upon that Ground, are past all Doubt. Here we examine the Thing in general, without taking notice of the Fallacy of our Senses upon some Occasions, and which Way Error is then to be avoided.

We cannot immediately judge of all Physical Matters by our Senses. We have then Recourse to another just Way of reasoning, though not Mathematical. It depends upon this Axiom; (viz.) *We must look upon as true, whatever being denied would destroy civil Society, and deprive us of the Means of Living.* From which Proposition the second and third Rules of the *Newtonian* Method most evidently follow.

For who could live a Minute's Time in Tranquility, if a Man was to doubt the Truth of what passes for certain, wherever Experiments have been made about it; and if he did not depend upon seeing the like Effects produced by the same Cause?

The following Reasonings, for Example, are daily taken for granted as undoubtedly true, without any previous Examination; because every Body sees that they cannot be called in Question without destroying the present Œconomy of Nature.

A Building, this Day firm in all its Parts, will not of itself run to Ruin to Morrow. Thus, by a Parity of Reason, the Cohesion and Gravity of the Parts of Bodies, which I never saw altered, nor heard of having been altered, without some intervening external Cause, will not be altered to Night, because the Cause of Cohesion and Gravity will be the same to Morrow as it is to Day. Who does not see, that the Certainty of this Reasoning depends only upon the Truth of the fore-mentioned Principle?

The Timber and Stones of any Country, which are fit for a Building, if brought over here, will serve in this Place, except what Changes may arise from an external Cause; and I shall no more fear the Fall of my Building, than the Inhabitants of the Country, from whence those Materials were brought, wou'd do, if they had built a House with them. Thus the Power which causes the Cohesion of Parts, and that which gives Weight to Bodies, are the same in all Countries.

I have used such Kind of Food for so many Years, therefore I will use it again to Day without Fear.

When I see Hemlock, I conclude it to be poisonous, tho' I never made an Experiment of that very Hemlock I see before my Eyes.

All these Reasonings are grounded upon Analogy; and there is no Doubt, but our Creator has in many Cases left us no other Way of Reasoning, and therefore it is a right Way. [...]

In Physics then we are to discover the Laws of Nature by the Phænomena, then by Induction prove them to be general Laws; all the rest is to be handled Mathematically. Whoever will seriously examine, what Foundation this Method Physics is built upon, will easily discover this to be the only true one, and that all Hypotheses are to be laid aside.

So much for the Method of philosophising. I have now a Word to say of the Work itself, of which this is the first Tome.

The whole Work is divided into four Books. The first treats of Body in general, and the Motion of Solid Bodies. The Second of Fluids. What belongs to Light is handled in the Third. The Fourth explains the motions of Celestial Bodies, and what has a Relation to them upon Earth. The two first Books are contained in this Tome.

In Order to render the Study of Natural Philosophy as easy and agreeable as possible, I have thought fit to illustrate every Thing by Experiments, and to set the very Mathematical Conclusions before the Reader's Eyes by this Method.

He, that sets forth the Elements of a Science, does not promise the learned World any Thing new in the main. Therefore I thought it needless, to point out where what is here contained is to be found. I have made my Property of whatever served my Purpose; and I thought giving Notice of it, once for all, was sufficient to avoid the Suspicion of Theft. I had rather lose the Honour of a few Discoveries, dispersed here and there in this Treatise, than rob any one of theirs. Let who will then take to himself what he thinks his own: I lay claim to nothing.

As to Machines which serve for making the Experiments, I have taken care to imitate several from other Authors, have altered and improved others, and added many new ones of my own Invention. And no Wonder I should be forced to that Necessity, having made Experiments upon many Things never tried perhaps by any one before. For Mathematicians think Experiments superfluous, where Mathematical Demonstrations will take Place: But as all Mathematical Demonstrations are abstracted, I do not question their becoming easier, when Experiments set forth the Conclusions before our Eyes; following therein the Example of the *English*, whose Way of teaching Natural Philosophy gave me Occasion to think of the Method I have followed in this Work. I shall always glory in treading in their Footsteps, who, with the Prince of Philosophers for their Guide, have first opened the Way to the Discovery of Truth in Philosophical Matters.

As to the Machines, I will say thus much more by Way of Advertisement, That most of them have been made by a very ingenious Artist of this Town, and no unskilful Philosopher, whose Name is *John van Musschenbroek*, and who has a perfect Knowledge of every Thing that is here explained. Which Advertisement, I suppose, will not be displeasing to those who may have a Fancy to get some of those Machines made for themselves.

Chapter Eleven
Science in the Scottish Enlightenment

11.1 James Hutton, *Abstract of a dissertation read in the Royal Society of Edinburgh ... concerning the system of the earth, its duration and stability*, 1785. Reprinted (intro. J. Craig) as *The 1785 Abstract* (Edinburgh: Scottish Academic Press, 1987), pp. 3–30

The purpose of this Dissertation is to form some estimate with regard to the time the globe of this Earth has existed, as a world maintaining plants and animals; to reason with regard to the changes which the earth has undergone; and to see how far an end or termination to this system of things may be perceived, from the consideration of that which has already come to pass.

As it is not in human record, but in natural history, that we are to look for the means of ascertaining what has already been, it is here proposed to examine the appearances of the earth, in order to be informed of operations which have been transacted in time past. It is thus that, from principles of natural philosophy, we may arrive at some knowledge of order and system in the economy of this globe, and may form a rational opinion with regard to the course of nature, or to events which are in time to happen.

The solid parts of the present land appear, in general, to have been composed of the productions of the sea, and of other materials similar to those now found upon the shores. Hence we find reason to conclude,

1*st*, That the land on which we rest is not simple and original, but that it is a composition, and had been formed by the operation of second causes.

2*dly*, That, before the present land was made, there had subsisted a world composed of sea and land, in which were tides and currents, with such operations at the bottom of the sea as now take place. And,

Lastly, That, while the present land was forming at the bottom of the ocean, the former land maintained plants and animals; at least, the sea was then inhabited by animals, in a similar manner as it is at present.

Hence we are led to conclude, that the greater part of our land, if not the whole, had been produced by operations natural to this globe; but that, in order to make this land a permanent body, resisting the operations of the waters, two things had been required; 1*st*, The consolidation of masses formed by collections of loose or incoherent materials; 2*dly*, The elevation of those consolidated masses from the bottom of the sea, the place where they were collected, to the stations in which they now remain above the level of the ocean. [. . .]

Thus the subject is considered as naturally divided into two branches, to be separately examined: *First*, by what natural operation strata of loose materials had been formed into solid masses; *Secondly*, By what power of nature the consolidated strata at the bottom of the sea had been transformed into land.

With regard to the *first* of these, the consolidation of strata, there are two ways in which this operation may be conceived to have been performed; first, by means of the solution of bodies in water, and the after concretion of these dissolved substances, when separated from their solvent; *secondly*, the fusion of bodies by means of heat, and the subsequent congelation of those consolidating substances. [. . .]

With regard to the other probable means, heat and fusion, these are found to be perfectly competent for producing the end in view, as every kind of substance may by heat be rendered soft, or brought into fusion, and as strata are actually found consolidated with every different species of substance. [. . .]

Having come to this general conclusion, that heat and fusion, not aqueous solution, had preceded the consolidation of the loose materials collected at the bottom of the sea, those consolidated strata, in general, are next examined, in order to discover other appearances, by which the doctrine may be either confirmed or refuted. Here the changes of strata, from their natural state of continuity, by veins and fissures, are considered; and the clearest evidence is hence deduced, that the strata have been consolidated by means of fusion, and not by aqueous solution; for, not only are strata in general found intersected with veins and cutters, an appearance inconsistent with their having been consolidated simply by previous solution; but, in proportion as strata are more or less consolidated, they are found with the proper corresponding appearances of veins and fissures.

With regard to the second branch, in considering by what power the consolidated strata had been transformed into land, or raised above the level of the sea, it is supposed that the same power of extreme heat, by which every different mineral substance had been brought into a melted state, might be capable of producing an expansive force, sufficient for elevating the land, from the bottom of the ocean, to the place it now occupies above the surface of the sea. Here we are again referred to nature, in examining how far the strata, formed by successive sediments or accumulations deposited at the bottom of the sea, are to be found in that regular state, which would necessarily take place in their original production; or if, on the other hand, they are actually changed in their natural situation, broken, twisted, and confounded, as might be expected, from the operation of subterranean heat, and violent expansion. But, as strata are actually found in every degree of fracture, flexure, and contortion, consistent with this supposition, and with no other, we are led to conclude, that our land had been raised above the surface of the sea, in order to become a habitable world; as well as that it had been consolidated by means of the same power of subterranean heat, in order to remain above the level of the sea, and to resist the violent efforts of the ocean.

This theory is next confirmed by the examination of mineral veins, those great fissures of the earth, which contain matter perfectly foreign to the strata they traverse; matter evidently derived from the mineral region, that is, from the place where the active power of fire, and the expansive force of heat, reside.

Such being considered as the operations of the mineral region, we are hence directed to look for the manifestation of this power and force, in the appearances of nature. It is here we find eruptions of ignited matter from the scattered volcano's of the globe; and these we conclude to be the effects of such a power precisely as that above which we now inquire. Volcano's are thus considered as the proper discharges of a superfluous or redundant power; not as things accidental in the course of nature, but as useful for the safety of mankind, and as forming a natural ingredient in the constitution of the globe.

Lastly, The extension of this theory, respecting mineral strata, to all parts of the globe, is made, by finding a perfect similarity in the solid land through all the earth, although, in particular places, it is attended with peculiar productions, with which the present inquiry is not concerned.

A theory is thus formed, with regard to a mineral system. In this system, hard and solid bodies are to be formed from soft bodies, from loose or incoherent materials, collected together at the bottom of the sea; and the bottom of the ocean is to be made to change its place with relation to the centre of the earth, to be formed into land above the level of the sea, and to become a country fertile and inhabited.

That there is nothing visionary in this theory, appears from its having been rationally deduced from natural events, from things which have already happened; things which have left, in the particular constitutions of bodies, proper traces of the manner of their production; and things which may be examined with all the accuracy, or reasoned upon with all the light, that science can afford. As it is only by employing science in this manner, that philosophy enlightens man with the knowledge of that wisdom or design which is to be found in nature, the system now proposed, from unquestionable principles, will claim the attention of scientific men, and may be admitted in our speculations with regard to the works of nature, notwithstanding many steps in the progress may remain unknown.

By thus proceeding upon investigated principles, we are led to conclude, that, if this part of the earth which we now inhabit had been produced, in the course of time, from the materials of a former earth, we should, in the examination of our land, find data from which to reason, with regard to the nature of that world, which had existed during the period of time in which the present earth was forming; and thus we might be brought to understand the nature of that earth which had preceded this; how far it had been similar to the present, in producing plants and nourishing animals. But this interesting point is perfectly ascertained, by finding abundance of every manner of vegetable production, as well as the several species of marine bodies, in the strata of our earth.

Having thus ascertained a regular system, in which the present land of the globe had been first formed at the bottom of the ocean, and then raised above the surface of the sea, a question naturally occurs with regard to time; what had been the space of time necessary for accomplishing this great work?

In order to form a judgement concerning this subject, our attention is directed to another progress in the system of the globe, namely, the destruction of the land which had preceded that on which we dwell. [...]

If we could measure the progress of the present land, towards its dissolution by attrition, and its submersion in the ocean, we might discover the actual duration of a former earth; an earth which had supported plants and animals, and had supplied the ocean with those materials which the construction of the present earth required; consequently, we should have the measure of a corresponding space of time, viz. that which had been required in the production of the present land. [...]

But, as there is not in human observation proper means for measuring the waste of land upon the globe, it is hence inferred, that we cannot estimate the duration of what we see at present, nor calculate the period at which it had begun; so that, with respect to human observation, this world has neither a beginning nor an end.

An endeavour is then made to support the theory by an argument of a moral nature, drawn from the consideration of a final cause. Here a comparison is formed between the present theory, and those by which there is necessarily implied either evil or disorder in natural things; and an argument is formed, upon the supposed wisdom of nature, for the justness of a theory in which perfect order is to be perceived. For,

According to the theory, a soil, adapted to the growth of plants, is necessarily prepared, and carefully preserved; and, in the necessary waste of land which is inhabited, the foundation is laid for future continents, in order to support the system of this living world.

Thus, either in supposing Nature wise and good, an argument is formed in confirmation of the theory, or, in supposing the theory to be just, an argument may be established for wisdom and benevolence to be perceived in nature. In this manner, there is opened to our view a subject interesting to man who thinks; a subject on which to reason with relation to the system of nature; and one which may afford the human mind both information and entertainment.

11.2 James Hutton, *Theory of the Earth*, vol. 1 of 2 vols, 1795, ed. Codicote, Wheldon and Wesley (J. Cramer, repr. 1972), pp. 175–83

Therefore, from the consideration of those materials which compose the present land, we have reason to conclude, that, during the time this land was forming, by the collection of its materials at the bottom of the sea, there had been a former land containing materials similar to those which we find at present in examining the earth. We may also conclude, that there had been operations similar to those which we now find natural to the globe, and necessarily exerted in the actual formation of gravel, sand, and clay. But what we have now chiefly in view to illustrate is this, that there had then been in the ocean a system of animated beings, which propagated their species, and which have thus continued their several races to this day.

In order to be convinced of that truth, we have but to examine the strata of our earth, in which we find the remains of animals. In this examination, we not only discover every genus of animal which at present exists in the sea, but probably every species, and perhaps some species with which at present we are not acquainted. There are, indeed, varieties in those species, compared with the present animals which we examine, but no greater varieties than may perhaps be found among the same species in the different quarters of the globe. Therefore, the system of animal life, which had been maintained in the ancient sea, had not been different from that which now subsists, and of which it belongs to naturalists to know the history.

It is the nature of animal life to be ultimately supported from matter of vegetable production. Inflammable matter may be considered as the *pabulum* of life. This is prepared in the bodies of living plants, particularly in their leaves exposed to the sun and light. This inflammable matter, on the contrary, is consumed in animal bodies, where it produces heat or light, or both. Therefore, however animal matter, or the pabulum of life, may circulate through a series of digesting powers, it is constantly impaired or diminishing in the course of this economy, and, without the productive power of plants, it would finally be extinguished*.

The animals of the former world must have been sustained during indefinite successions of ages. The mean quantity of animal matter, therefore, must have been preserved by vegetable production, and the natural waste of inflammable substance repaired with continual addition; that is to say, the quantity of inflammable matter necessary to the animal consumption, must have been provided by means of vegetation. Hence we must conclude, that there had been a world of plants, as well as an ocean replenished with living animals.

We are now, in reasoning from principles, come to a point decisive of the question, and which will either confirm the theory, if it be just, or confute

---

* See Dissertations on different subjects of Natural Philosophy, part II.

our reasoning, if we have erred. Let us, therefore, open the book of Nature, and read in her records, if there had been a world bearing plants, at the time when this present world was forming at the bottom of the sea.

Here the cabinets of the curious are to be examined; but here some caution is required, in order to distinguish things perfectly different, which sometimes are confounded.

Fossil wood, to naturalists in general, is wood dug up from under ground, without inquiring whether this had been the production of the present earth, or that which had preceded it in the circulation of land and water. The question is important, and the solution of it is, in general, easy. The vegetable productions of the present earth, however deep they may be found buried beneath its surface, and however ancient they may appear, compared with the records of our known times, are new, compared with the solid land on which they grew; and they are only covered with the produce of a vegetable soil, or the alluvion of the present land on which we dwell, and on which they had grown. But the fossil bodies which form the present subject of inquiry, belonged to former land, and are found only in the sea-born strata of our present earth. It is to these alone that we appeal, in order to prove the certainty of former events.

Mineralised wood, therefore, is the object now inquired after; that wood which had been lodged in the bottom of the sea, and there composed part of a stratum, which hitherto we have considered as only formed of the materials proper to the ocean. Now, what a profusion of this species of fossil wood is to be found in the cabinets of collectors, and even in the hands of lapidaries, and such artificers of polished stones! In some places, it would seem to be as common as the agate.

I shall only mention a specimen in my own collection. It is wood petrified with calcareous earth, and mineralised with pyrites. This specimen of wood contains in itself, even without the stratum of stone in which it is embedded, the most perfect record of its genealogy. It had been eaten or perforated by those sea worms which destroy the bottoms of our ships. There is the clearest evidence of this truth. Therefore, this wood had grown upon land which stood above the level of the sea, while the present land was only forming at the bottom of the ocean.

Wood is the most substantial part of plants, as shells are the more permanent part of marine animals. It is not, however, the woody part alone of the ancient vegetable world that is transmitted to us in the record of our mineral pages. We have the type of many species of foliage, and even of the most delicate flower; for, in this way, naturalists have determined, according to the Linnaean system, the species, or at least the genus, of the plant. Thus, the existence of a vegetable system at the period now in contemplation, so far from being doubtful, is a matter of physical demonstration.

The profusion of this vegetable matter, delivered into the ocean, which then generated land, is also evidenced in the amazing quantities of mineral coal which is to be found in perhaps every region of the earth.

Nothing can be more certain, than that all the coaly or bituminous strata have had their origin from the substance of vegetable bodies that grew upon the land. Those strata, tho', in general, perfectly consolidated, often separate horizontally in certain places; and there we find the fibrous or vascular structure of the vegetable bodies. Consequently, there is no doubt of fossil coal being a substance of vegetable production, however animal substances also may have contributed in forming this collection of oleaginous or inflammable matter.

Having thus ascertained the state of a former earth, in which plants and animals had lived, as well as the gradual production of the present earth, composed from the materials of a former world, it must be evident, that here are two operations which are necessarily consecutive. The formation of the present earth necessarily involves the destruction of continents in the ancient world; and, by pursuing in our mind the natural operations of a former earth, we clearly see the origin of that land, by the fertility of which, we, and all the animated bodies of the sea, are fed. It is in like manner, that, contemplating the present operations of the globe, we may perceive the actual existence of those productive causes, which are now laying the foundation of land in the unfathomable regions of the sea, and which will, in time, give birth to future continents.

But though, in generalising the operations of nature, we have arrived at those great events, which, at first sight, may fill the mind with wonder and with doubt, we are not to suppose, that there is any violent exertion of power, such as is required in order to produce a great event in little time; in nature, we find no deficiency in respect of time, nor any limitation with regard to power. But time is not made to flow in vain; nor does there ever appear the exertion of superfluous power, or the manifestation of design, not calculated in wisdom to effect some general end.

The events now under consideration may be examined with a view to see this truth; for it may be inquired, Why destroy one continent in order to erect another? The answer is plain; Nature does not destroy a continent from having wearied of a subject which had given pleasure, or changed her purpose, whether for a better or a worse; neither does she erect a continent of land among the clouds, to shew her power, or to amaze the vulgar man; Nature has contrived the productions of vegetable bodies, and the sustenance of animal life, to depend upon the gradual but sure destruction of a continent; that is to say, these two operations necessarily go hand in hand. But with such wisdom has nature ordered things in the economy of this world, that the destruction of one continent is not brought about without the renovation of the earth in the production of another; and the animal and vegetable bodies, for which the world above the surface of the sea is levelled with its bottom, are among the means employed in those operations, as well as the sustenance of those living beings is the proper end in view.

Thus, in understanding the proper constitution of the present earth, we are led to know the source from whence had come all the materials which nature had employed in the construction of the world which appears; a world contrived in consummate wisdom for the growth and habitation of a great diversity of plants and animals; and a world peculiarly adapted to the purposes of man, who inhabits all its climates, who measures its extent, and determines its productions at his pleasure.

11.3 Joseph Black, 'Experiments upon Magnesia Alba, Quicklime and some other Alkaline Substances', *Essays and Observations, Physical and Literary*, Edinburgh, 1756, pp. 157–225, in Alembic Club Reprint no. 1 (Edinburgh: 1944), pp. 10–20

MAGNESIA is quickly dissolved with violent effervescence, or explosion of air, by the acids of vitriol, nitre, and of common salt, and by distilled vinegar; the neutral saline liquors thence produced having each their peculiar properties.

THAT which is made with the vitriolic acid, may be condensed into crystals similar in all respects to epsom-salt.

THAT which is made with the nitrous is of a yellow colour, and yields saline crystals, which retain their form in a very dry air, but melt in a moist one.

THAT which is produced by means of spirit of salt, yields no crystals; and if evaporated to dryness, soon melts again when exposed to the air.

THAT which is obtained from the union of distilled vinegar with *magnesia*, affords no crystals by evaporation, but is condensed into a saline mass, which, while warm, is extremely tough and viscid, very much resembling a strong glue both in colour and consistence, and becomes brittle when cold.

By these experiments *magnesia* appears to be a substance very different from those of the calcarious class; under which I would be understood to comprehend all those that are converted into a perfect quick-lime in a strong fire, such as *lime-stone, marble, chalk,* those *spars* and *marles* which effervesce with aqua fortis, all *animal shells* and the bodies called *lithophyta*. All of these, by being joined with acids, yield a set of compounds which are very different from those we have just now described. Thus, if a small quantity of any calcarious matter be reduced to a fine powder and thrown into spirit of vitriol, it is attacked by this acid with a brisk effervescence; but little or no dissolution ensues. It absorbs the acid, and remains united with it in the form of a white powder, at the bottom of the vessel; while the liquor has hardly any taste, and shews only a very light cloud upon the addition of alkali.*

THE same white powder is also formed when spirit of vitriol is added to a calcarious earth dissolved in any other acid; the vitriolic expelling the other acid, and joining itself to the earth by a stronger attraction; [...]

THREE drams of *magnesia* in fine powder, an ounce of salt ammoniac, and six ounces of water were mixed together, and digested six days in a retort joined to a receiver.

DURING the whole time, the neck of the retort was pointed a little upwards, and the most watery part of the vapour, which was condensed there, fell back into its body. In the beginning of the experiment, a volatile salt was therefore collected in a dry form in the receiver, and afterwards dissolved into spirit.

WHEN all was cool, I found in the retort a saline liquor, some undissolved *magnesia*, and some salt ammoniac crystallized. The saline liquor was separated from the other two, and then mixed with the alkaline spirit. A coagulum was immediately formed, and a *magnesia* precipitated from the mixture.

THE *magnesia* which had remained in the retort, when well washed and dried, weighed two scruples and fifteen grains.

WE learn by the latter part of this experiment, that the attraction of the volatile alkali for acids is stronger than that of *magnesia*, since it separated this powder from the acid to which it was joined. But it also appears, that a gentle heat is capable of overcoming this superiority of attraction, and of gradually elevating the alkali, while it leaves the less volatile acid with the *magnesia*.

DISSOLVE a dram of any calcarious substance in the acid of nitre or of common salt, taking care that the solution be rendered perfectly neutral, or that no superfluous acid be added. Mix with this solution a dram of *magnesia* in fine powder, and digest it in the heat of boiling water about twenty four hours; then dilute the mixture with double its quantity of water, and filtrate. The greatest part of the earth now left in the filtre is calcarious, and the liquor which passed thro', if mixed with a dissolved alkali; yields a white powder, the largest portion of which is a true *magnesia*.

FROM this experiment it appears, that an acid quits a calcarious earth to join itself to *magnesia*; but the exchange being performed slowly, some of the

* Mr. *Margraaf* has lately demonstrated, by a set of curious and accurate experiments, that this powder is of the nature, and possesses the properties, of the gypseous or selenitic substances. That such substances can be resolved into vitriolic acid and calcarious earth, and can be again composed by joining these two ingredients together. Mem. de l'Acad. de Berlin. an. 1750, p. 144.

magnesia is still undissolved, and part of the calcarious earth remains yet joined to the acid. [...]

QUICK-LIME itself is also rendered mild by *magnesia*, if these two are well rubbed together and infused with a small quantity of water.

BY the following experiments, I proposed to know whether this substance could be reduced to a quick-lime.

AN ounce of *magnesia* was exposed in a crucible for about an hour to such a heat as is sufficient to melt copper. When taken out, it weighed three drams and one scruple, or had lost $\frac{7}{12}$ of its former weight.

I repeated, with the *magnesia* prepared in this manner, most of those experiments I had already made upon it before calcination, and the result was as follows.

IT dissolves in all the acids, and with these composes salts exactly similar to those described in the first set of experiments; but what is particularly to be remarked, it is dissolved without any the least degree of effervescence.

IT slowly precipitates the corrosive sublimate of mercury in the form of a black powder.

IT separates the volatile alkali in salt ammoniac from the acid, when it is mixed with a warm solution of that salt. But it does not separate an acid from a calcarious earth, nor does it induce the least change upon lime-water.

LASTLY, when a dram of it is digested with an ounce of water in a bottle for some hours, it does not make any the least change in the water. The *magnesia*, when dried, is found to have gained ten grains; but it neither effervesces with acids, nor does it sensibly affect lime-water.

OBSERVING *magnesia* to lose such a remarkable proportion of its weight in the fire, my next attempts were directed to the investigation of this volatile part, and, among other experiments, the following seemed to throw some light upon it.

THREE ounces of *magnesia* were distilled in a glass retort and receiver, the fire being gradually increased until the *magnesia* was obscurely red hot. When all was cool, I found only five drams of a whitish water in the receiver, which had a faint smell of the spirit of hartshorn, gave a green colour to the juice of violets, and rendered the solutions of corrosive sublimate and of silver very slightly turbid. But it did not sensibly effervesce with acids.

THE *magnesia* when taken out of the retort, weighed an ounce, three drams, and thirty grains, or had lost more than the half of its weight. It still effervesced pretty briskly with acids, tho' not so strongly as before this operation.

THE fire should have been raised here to the degree requisite for the perfect calcination of *magnesia*. But even from this imperfect experiment, it is evident, that of the volatile parts contained in that powder, a small proportion only is water; the rest cannot, it seems, be retained in vessels, under a visible form. Chemists have often observed, in their distillations, that part of a body has vanished from their senses, notwithstanding the utmost care to retain it; and they have always found, upon further inquiry, that subtile part to be air, which having been imprisoned in the body, under a solid form was set free and rendered fluid and elastic by the fire. We may therefore safely conclude, that the volatile matter, lost in the calcination of *magnesia*, is mostly air; and hence the calcined *magnesia* does not emit air, or make an effervescence, when mixed with acids.

THE water, from its properties, seems to contain a small portion of volatile alkali, which was probably formed from the earth, air, and water, or from some of these combined together; and perhaps also from a small quantity of inflammable matter which adhered accidentally to the *magnesia*. Whenever Chemists meet with this salt, they are inclined to ascribe its origin to some animal, or putrid vegetable, substance; and this they have always done, when they obtained it from the calcarious earths, all of which afford a small quantity of it. There is, however, no doubt that it can sometimes be produced independently of any such mixture, since many fresh vegetables and tartar afford a considerable quantity of it. And how can it, in the present instance, be supposed, that any animal or vegetable matter adhered to the *magnesia*, while it was dissolved by an acid, separated from this by an alkali, and washed with so much water?

TWO drams of *magnesia* were calcined in a crucible, in the manner described above, and thus reduced to two scruples and twelve grains. This calcined *magnesia* was dissolved in a sufficient quantity of spirit of vitriol, and then again separated from the acid by the addition of an alkali, of which a large quantity is necessary for this purpose. The *magnesia* being very well washed and dryed, weighed one dram and fifty grains. It effervesced violently, or emitted a large quantity of air, when thrown into acids, formed a red powder when mixed with a solution of sublimate, separated the calcarious earths from an acid, and sweetened lime-water: and had thus recovered all those properties which it had but just now lost by calcination: nor had it only recovered its original properties, but acquired besides an addition of weight nearly equal to what had been lost in the fire; and, as it is found to effervesce with acids, part of the addition must certainly be air.

THIS air seems to have been furnished by the alkali from which it was separated by the acid; for Dr. *Hales* has clearly proved, that alkaline salts contain a large quantity of fixed air, which they emit in great abundance when joined to a pure acid. In the present case, the alkali is really joined to an acid, but without any visible emission of air; and yet the air is not retained in it: for the neutral salt, into which it is converted, is the same in quantity, and in every other respect, as if the acid employed had not been previously saturated with *magnesia*, but offered to the alkali in its pure state, and had driven the air out of it in their conflict. It seems therefore evident, that the air was forced from the alkali by the acid, and lodged itself in the *magnesia*.

THESE considerations led me to try a few experiments, whereby I might know what quantity of air is expelled from an alkali, or from *magnesia*, by acids.

TWO drams of a pure fixed alkaline salt, and an ounce of water, were put into a Florentine flask, which, together with its contents, weighed two ounces and two drams. Some oil of vitriol diluted with water was dropt in, until the salt was exactly saturated; which it was found to be, when two drams, two scruples, and three grains of this acid had been added. The vial with its contents now weighed two ounces, four drams, and fifteen grains. One scruple, therefore, and eight grains were lost during the ebullition, of which a trifling portion may be water, or something of the same kind. The rest is air. [...]

THE celebrated *Homberg* has attempted to estimate the quantity of solid salt contained in a determined portion of the several acids. He saturated equal quantities of an alkali with each of them; and, observing the weight which the alkali had gained, after being perfectly dryed, took this for the quantity of solid salt contained in that share of the acid which performed the saturation. But we learn from the above experiment, that his estimate was not accurate, because the alkali loses weight as well as gains it.

TWO drams of *magnesia*, treated exactly as the alkali in the last experiment, were just dissolved by four drams, one scruple, and seven grains of the same acid liquor, and lost one scruple and sixteen grains by the ebullition.

TWO drams of *magnesia* were reduced, by the action of a violent fire, to two scruples and twelve grains, with which the same process was repeated, as in the two last experiments; four drams, one scruple, and two grains of the same acid were required to compleat the solution, and no weight was lost in the experiment.

AS in the separation of the volatile from the fixed parts of bodies, by means of heat, a small quantity of the latter is generally raised with the former; so the air and water, originally contained in the *magnesia*, and afterwards dissipated by the fire, seem to have carried off a small part of the fixed earth of this substance. This is probably the reason, why calcined *magnesia* is saturated with a quantity of acid, somewhat less than what is required to dissolve it before calcination; and the same may be assigned as one cause which hinders us from restoring the whole of its original weight, by solution and precipitation.

I took care to dilute the vitriolic acid, in order to avoid the heat and ebullition which it would otherwise have excited in the water; and I chose a

Florentine flask, on account of its lightness, capacity, and shape, which is peculiarly adapted to the experiment; for the vapours raised by the ebullition circulated for a short time, thro' the wide cavity of the vial, but were soon collected upon its sides, like dew, and none of them seemed to reach the neck, which continued perfectly dry to the end of the experiment.

WE now perceive the reason, why crude and calcined *magnesia*, which differ in many respects from one another, agree however in composing the same kind of salt, when dissolved in any particular acid; for the crude *magnesia* seems to differ from the calcined chiefly by containing a considerable quantity of air, which air is unavoidably dissipated and lost during the dissolution.

FROM our experiments, it seems probable, that the increase of weight which some metals acquire, by being first dissolved in acids, and then separated from them again by alkalis, proceeds from air furnished by the alkalis. [...]

THE above experiments lead us also to conclude, that volatile alkalis, and the common absorbent earths, which lose their air by being joined to acids, but shew evident signs of their having recovered it, when separated from them by alkalis, received it from these alkalis which lost it in the instant of their joining with the acid.

11.4 Adam Smith, *Essays on Philosophical Subjects*, 1795, ed. W. P. D. Wightman and J. C. Bryce (Oxford: Clarendon Press, 1980), pp. 40–2, 45–6

When one accustomed object appears after another, which it does not usually follow, it first excites, by its unexpectedness, the sentiment properly called Surprise, and afterwards, by the singularity of the succession, or order of its appearance, the sentiment properly called Wonder. We start and are surprised at feeling it there, and then wonder how it came there. The motion of a small piece of iron along a plain table is in itself no extraordinary object, yet the person who first saw it begin, without any visible impulse, in consequence of the motion of a loadstone at some little distance from it could not behold it without the most extreme Surprise; and when that momentary emotion was over, he would still wonder how it came to be conjoined to an event with which, according to the ordinary train of things, he could have so little suspected it to have any connection.

When two objects, however unlike, have often been observed to follow each other, and have constantly presented themselves to the senses in that order, they come to be so connected together in the fancy, that the idea of the one seems, of its own accord, to call up and introduce that of the other. If the objects are still observed to succeed each other as before, this connection, or, as it has been called, this association of their ideas, becomes stricter and stricter, and the habit of the imagination to pass from the conception of the one to that of the other, grows more and more rivetted and confirmed. As its ideas move more rapidly than external objects, it is continually running before them, and therefore anticipates, before it happens, every event which falls out according to this ordinary course of things. When objects succeed each other in the same train in which the ideas of the imagination have thus been accustomed to move, and in which, though not conducted by that chain of events presented to the senses, they have acquired a tendency to go on of their own accord, such objects appear all closely connected with one another, and the thought glides easily along them, without effort and without interruption. They fall in with the natural career of the imagination; and as the ideas which represented such a train of things would seem all mutually to introduce each other, every last thought to be called up by the foregoing, and to call up the succeeding; so when the objects themselves occur, every last event seems, in the same manner, to be introduced by the foregoing, and to introduce the succeeding. There is no break, no stop, no gap, no interval. The ideas excited by so coherent a chain of things seem, as it were, to float through the mind of their own accord, without obliging it to exert itself, or to make any effort in order to pass from one to them to another.

But if this customary connection be interrupted, if one or more objects appear in an order quite different from that to which the imagination has been accustomed, and for which it is prepared, the contrary of all this happens. We are at first surprised by the unexpectedness of the new appearance, and when that momentary emotion is over, we still wonder how it came to occur in that place. The imagination no longer feels the usual facility of passing from the event which goes before to that which comes after. It is an order or law of succession to which it has not been accustomed, and which it therefore finds some difficulty in following, or in attending to. The fancy is stopped and interrupted in that natural movement or career, according to which it was proceeding. Those two events seem to stand at a distance from each other; it endeavours to bring them together, but they refuse to unite; and it feels, or imagines it feels, something like a gap or interval betwixt them. It naturally hesitates, and, as it were, pauses upon the brink of this interval; it endeavours to find out something which may fill up the gap, which, like a bridge, may so far at least unite those seemingly distant objects, as to render the passage of the thought betwixt them smooth, and natural, and easy. The supposition of a chain of intermediate, though invisible, events, which succeed each other in a train similar to that in which the imagination has been accustomed to move, and which link together those two disjointed appearances, is the only means by which the imagination can fill up this interval, is the only bridge which, if one may say so, can smooth its passage from the one object to the other. Thus, when we observe the motion of the iron, in consequence of that of the loadstone, we gaze and hesitate, and feel a want of connection betwixt two events which follow one another in so unusual a train. But when, with Des

Cartes, we imagine certain invisible effluvia to circulate round one of them, and by their repeated impulses to impel the other, both to move towards it, and to follow its motion, we fill up the interval betwixt them, we join them together by a sort of bridge, and thus take off that hesitation and difficulty which the imagination felt in passing from the one to the other. That the iron should move after the loadstone seems, upon this hypothesis, in some measure according to the ordinary course of things. Motion after impulse is an order of succession with which of all things we are the most familiar. Two objects which are so connected seem no longer to be disjoined, and the imagination flows smoothly and easily along them. [...]

Philosophy is the science of the connecting principles of nature. Nature, after the largest experience that common observation can acquire, seems to abound with events which appear solitary and incoherent with all that go before them, which therefore disturb the easy movement of the imagination; which make its ideas succeed each other, if one may say so, by irregular starts and sallies; and which thus tend, in some measure, to introduce those confusions and distractions we formerly mentioned. Philosophy, by representing the invisible chains which bind together all these disjointed objects, endeavours to introduce order into this chaos of jarring and discordant appearances, to allay this tumult of the imagination, and to restore it, when it surveys the great revolutions of the universe, to that tone of tranquillity and composure, which is both most agreeable in itself, and most suitable to its nature.

Chapter Twelve
Science on the Fringe of Europe: Eighteenth-Century Sweden

12.1 Carl Linnaeus, Dedication and Preface, *Species Plantarum*, 1753, trans. W. T. Stearn (London: Ray Society, 1957), pp. 152–5

MOST MIGHTY ALL GRACIOUS

KING AND QUEEN

ADOLPH FRIDERIC,

LOVISA ULRICA,

OF THE SWEDES GOTHS AND WENDS

KING AND QUEEN!

TO YOUR MAJESTIES I most humbly owe this work, the fruit of the greatest and best part of my life, which under YOUR MAJESTIES' genial reign I now complete with a contented and tranquil mind.

YOUR MAJESTIES, from your first happy arrival in this kingdom, have looked upon me with kindly eyes, called me to Your Court and expressed towards me all Royal grace.

YOUR MAJESTIES have also given the Sciences which I follow light and prestige in Your Realm, since YOUR MAJESTIES have not only most graciously had the Creator's wondrous work described for you but have also assembled representative products of all three realms of Nature and have commanded them to be installed within your splendid palaces at Ulriksdahl and Drottningholm where, as representatives of the whole wide world, they can daily come before the eyes of YOUR MAJESTIES.

Believe me that I do not say too much when I openly admit that from the time of the wise King Solomon and the Queen of rich Arabia no Sovereigns have shown greater respect for the wonders of the Almighty Creator.

For this Science ought and shall vie with all other monuments to make YOUR MAJESTIES' names undying, by imprinting them in Nature's own book, which every year is issued anew, being as lasting as the Terrestrial Globe.

May Heaven make YOUR MAJESTIES' days and reign long and prosperous, as surely as I live and die YOUR REGAL MAJESTIES'

most humble and most faithful
servant and subject
CARL LINNAEUS

To the Well-disposed Reader

MAN as a sentient being contemplates the WORLD as the theatre of the Almighty, everywhere adorned with the greatest wonders of All-knowing wisdom, and he has been brought into it as a kind of Guest so that, whilst taking pleasure in these delights, he should acknowledge the greatness of the Lord. Unworthy to be accepted as guest would be one who, in the manner of cattle, attended only to his gullet and knew not how to admire and value the mighty works of his host.

In order to go forth as GUESTS worthy of our world, we must examine carefully these works of the Creator, which the supreme Being has in such a manner bound up with our well-being that we need not lack any good things, and the more we understand these, the more they yield for the use of humanity.

Properly to acquire a KNOWLEDGE of these, it is necessary to link together a single distinct *concept* and a distinct *name*, for by neglecting this the abundance of objects would overwhelm us and all exchange of information would cease through lack of a common language.

Hence NATURAL SCIENCE was born among men, embracing under *Physics* and *Chemistry* the elements, under *Zoology, Botany* and *Mineralogy* the three realms of Nature.

It is my pleasure here to take BOTANY as my special study, which previously was the knowledge of a few plants; to-day however the abundance of material for choice has made it the most extensive of all [the sciences].

To the unwearied efforts of DISCOVERERS in recent times, among whom are especially to be named *Clusius (L'Ecluse), Columna (Colonna), the Bauhins, Hermann, Rheede, Sherard, Ray, Plukenet, Tournefort, Plumier, Vaillant, Dillenius, Gmelin* etc., we own our knowledge of many more plants than to the learned of ancient times.

KNOWLEDGE of plants formerly consisted of [knowing] arbitrary Names impressed on the tablets of *memory*, the waverings of which were mitigated by illustrations.

The wisdom of SYSTEMATISTS searched out an arrangement to help the memory and built the science on solid foundations, which we owe to the Outstanding Men *Gesner, Cesalpino, Bauhin, Morison, Hermann, Tournefort, Vaillant, Dillenius* and others.

The ARIADNEAN thread of the Systematists ended with the *Genera*. I have tried to extend it to the *Species*, having devised proper differential characters, so that certainty even in these might be established, since *all true knowledge rests upon a knowledge of the Species;* and if this is lacking, the record becomes uncertain, as in several narratives by *Travellers*.

To become well acquainted with the SPECIES of plants, I have travelled through the mountains (Alpes) of *Lapland*, all *Sweden*; parts of *Norway, Denmark, Germany*, the *Netherlands, England, France*; have after that diligently searched the botanic GARDENS of *Paris, Oxford, Chelsea, Hartekamp, Leyden, Utrecht, Amsterdam, Uppsala* and other places; have also consulted the HERBARIA of *Burser, Hermann, Clifford, Burman, Gronovius, Royen, Sloane, Sherard, Bobart, Miller, Surian, Tournefort, Vaillant, Jussieu, Bäck* and others. At my instigation my most cherished one-time STUDENTS have gone abroad, *P. Kalm* to Canada, *F. Hasselquist* to Egypt, *P. Osbeck* to China, *P. Löfling* to Spain, *Montin* to Lapland, and have sent me the plants gathered. Moreover from various countries my BOTANICAL FRIENDS have sent me not a few seeds and dried plants, notably *B. Jussieu, Royen, J. Gesner, Wachendorf, Sibthorp, Monti, Gleditsch, Krascheninnikow, Minuart, Velez*, as well as Baron *O. Munchhausen*, Baron *S. C. Bielke*, Ritter *J. Rathgeb*, Nobleman *Demidoff, Collinson, Torèn, Braad* and others. *Clifford* gave me all that he had in duplicate; *Lagerström* many from the East Indies; *Gronovius* many Virginian and *Gmelin* pretty well all Siberian, and *Sauvages* his entire collection of plants, a rare and unheard of act, whereby I have acquired no ordinary wealth of plants.

Before now I have set out the specific DIFFERENTIAL CHARACTERS for not a few plants in the *Flora Lapponica, Flora Suecica* and *Flora Zeylanica* and in the *Hortus Cliffortianus* and *Hortus Upsaliensis*. The outstanding Botanists *Gronovius, Royen, Wachendorf, Gorter, B. Jussieu, Le Monnier, Guettard, Dalibard, Sauvages, Colden* and *Hill* have adopted the same principles, as have to some degree *Haller, Gmelin* and others, through whose works many species have been made evident and settled.

These widely scattered NAMES I have planned to bring together for the sake of students, adding plants since acquired, and reducing all to one system; through many more *Species* being observed, more outstanding *marked characters* being detected and more apt *Terms* being coined, I have had sometimes to amend the differential characters, excellent though they were hitherto.

To determine the essential characters for the SPECIFIC NAME is no light task; indeed it demands an accurate knowledge of many *Species*, a most scrupulous investigation of their *parts*, the choice of *different characters*, and then the proper use of *Terminology*, that they may be expressed most concisely and surely.

I have omitted here the PLANTS NOT SEEN, being many times deceived by authors, so as not to mix the doubtful ones with those quite certain; if indeed on occasion I have been unable sufficiently to examine a plant or have obtained an imperfect specimen, I have marked this with the sign †, that others may examine the same more accurately.

Plants not named in this little book (Libellus), if any one sends them to me, I shall list in the next edition, God willing, with honorable mention of the giver.

That the NUMBER of plants of the whole world is much less than is commonly believed I ascertained by fairly safe calculation, in as much as it hardly reaches 10,000.

I have put TRIVIAL names in the margin so that without more ado we can represent one plant by one name; these I have taken, it is true, without special choice, leaving this for another day. However, I would warn most solemnly all sensible botanists not to propose a trivial name without adequate specific distinction, lest the science falls back into its early crude state.

For EUROPEAN plants I have included very few SYNONYMS, being content with *C. Bauhin* and an outstanding *Illustrator*; for EXOTICS, however, several, because they are more difficult and less familiar.

Only in doubtful instances was it necessary to add DESCRIPTIONS, and these without uncertainties, in order to keep the handbook suitable for beginners.

The original PLACE of growth I have added in my usual manner, and for the best known I have indicated *Shrubby Plants* by ♄, *Perennials* ♃, *Biennials* ♂, *Annuals* ☉.

I have employed (or added) some *new* GENERA, some *unchanged*, which it is my intention to put forward very soon in the new edition of the Genera plantarum.

The darts of ADVERSARIES I have never cast back; with an undisturbed mind I endured the most bitter abuse, accusations, jeers and trumpeting (in every age the rewards of labour for the most outstanding Men), and I do not envy their authors if they thereby receive greater glory from the crowd. I put up with this without turning a hair, and why should I not bear with it from the unjust, I who am overladen with the highest praises by genuine and indeed most accomplished Botanists to whom my adversaries must defer. Neither my heavy-growing years nor the position I hold, nor my character allow me to give my adversaries like for like. What of brief life remains for me I shall devote in calmness to more useful observations. The things of nature truly follow their own laws, so that just as errors regarding them cannot be defended by anybody, so truths resting upon observations cannot be trampled under foot even by the whole world of the learned; therefore I call upon our grandchildren to be the judges:

Envy feeds upon the living, after death is stilled,
Then honour watches over a man according to his merit.

Written the 2nd of May, 1753 at Uppsala.

12.2 C. W. Scheele, *Chemical Investigations on Air and Fire*, 1777, trans. L. Dobbin (Edinburgh: Alembic Club Reprint no. 8, 1952), pp. 5–15

1. It is the object and chief business of chemistry to skilfully separate substances into their constituents, to discover their properties, and to compound them in different ways.

How difficult it is, however, to carry out such operations with the greatest accuracy, can only be unknown to one who either has never undertaken this occupation, or at least has not done so with sufficient attention.

2. Hitherto chemical investigators are not agreed as to how many elements or fundamental materials compose all substances. In fact this is one of the most difficult problems; some indeed hold that there remains no further hope of searching out the elements of substances. Poor comfort for those who feel their greatest pleasure in the investigation of natural things! Far is he mistaken, who endeavours to confine chemistry, this noble science, within such narrow bounds! Others believe that earth and phlogiston are the things from which all material nature has derived its origin. The majority seem completely attached to the peripatetic elements.

3. I must admit that I have bestowed no little trouble upon this matter in order to obtain a clear conception of it. One may reasonably be amazed at the numerous ideas and conjectures which authors have recorded on the subject, especially when they give a decision respecting the fiery phenomenon; and this very matter was of the greatest importance to me. I perceived the necessity of a knowledge of fire, because without this it is not possible to make any experiment; and without fire and heat it is not possible to make use of the action of any solvent. I began accordingly to put aside all explanations of fire; I undertook a multitude of experiments in order to fathom this beautiful phenomenon as fully as possible. I soon found, however, that one could not form any true judgment regarding the phenomena which fire presents, without a knowledge of the air. I saw, after carrying out a series of experiments, that air really enters into the mixture of fire, and with it forms a constituent of flame and of sparks. I learned accordingly that a treatise like this, on fire, could not be drawn up with proper completeness without taking the air also into consideration.

4. Air is that fluid invisible substance which we continually breathe, which surrounds the whole surface of the earth, is very elastic, and possesses weight. It is always filled with an astonishing quantity of all kinds of exhalations, which are so finely subdivided in that they are scarcely visible even in the sun's rays. Water vapours always have the preponderance amongst these foreign particles. The air, however, is also mixed with another elastic substance resembling air, which differs from it in numerous properties, and is, with good reason, called aerial acid by Professor Bergman. It owes its presence to organised bodies, destroyed by putrefaction or combustion.

5. Nothing has given philosophers more trouble for some years than just this delicate acid or so-called fixed air. Indeed it is not surprising that the conclusions which one draws from the properties of this elastic acid are not favourable to all who are prejudiced by previously conceived opinions. These defenders of the Paracelsian doctrine believe that the air is in itself unalterable; and, with Hales, that it really unites with substances thereby losing its elasticity; but that it regains its original nature as soon as it is driven out of these by fire or fermentation. But since they see that the air so produced is endowed with properties quite different from common air, they conclude, without experimental proofs, that this air has united with foreign materials, and that it must be purified from these admixed foreign particles by agitation and filtration with various liquids. I believe that there would be no hesitation in accepting this opinion, if one could only demonstrate clearly by experiments that a given quantity of air is capable of being completely converted into fixed or other kind of air by the admixture of foreign materials; but since this has not been done, I hope I do not err if I assume as many kinds of air as experiment reveals to me. For when I have collected an elastic fluid, and observe concerning it that its expansive power is increased by heat and diminished by cold, while it still uniformly retains its elastic fluidity, but also discover in it properties and behaviour different from those of common air, then I consider myself justified in believing that this is a peculiar kind of air. I say that air thus collected must retain its elasticity even in the greatest cold, because otherwise an innumerable multitude of varieties of air would have to be assumed, since it is very probable that all substances can be converted by excessive heat into a vapour resembling air.

6. Substances which are subjected to putrefaction or to destruction by means of fire diminish, and at the same time consume, a part of the air; sometimes it happens that they perceptibly increase the bulk of the air, and sometimes finally that they neither increase nor diminish a given quantity of air – phenomena which are certainly remarkable. Conjectures can here determine nothing with certainty, at least they can only bring small satisfaction to a chemical philosopher, who must have his proofs in his hands. Who does not see the necessity of making experiments in this case, in order to obtain light concerning this secret of nature?

7. General properties of ordinary air
(1.) Fire must burn for a certain time in a given quantity of air. (2.) If, so far as can be seen, this fire does not produce during combustion any fluid resembling air, then, after the fire has gone out of itself, the quantity of air must be diminished between a third and a fourth part. (3.) It must not unite

with common water. (4.) All kinds of animals must live for a certain time in a confined quantity of air. (5.) Seeds, as for example peas, in a given quantity of similarly confined air, must strike roots and attain a certain height with the aid of some water and of a moderate heat.

Consequently, when I have a fluid resembling air in its external appearance, and find that it has not the properties mentioned, even when only one of them is wanting, I feel convinced that it is not ordinary air. [...]

19. Third Experiment. – I placed 3 teaspoonfuls of iron filings in a bottle capable of holding 2 ounces of water; to this I added an ounce of water, and gradually mixed with them half an ounce of oil of vitriol. A violent heating and fermentation took place. When the froth had somewhat subsided, I fixed into the bottle an accurately fitting cork, through which I had previously fixed a glass tube A (Fig. 1). I placed this bottle in a vessel filled with hot water, B B (cold water would greatly retard the solution). I then approached a burning candle to the orifice of the tube, whereupon the inflammable air took fire and burned with a small yellowish-green flame. As soon as this had taken place, I took a small flask C, which was capable of holding 20 ounces of water, and held it so deep in the water that the little flame stood in the middle of the flask. The water at once began to rise gradually into the flask, and when the level had reached the point D the flame went out. Immediately afterwards the water began to sink again, and was entirely driven out of the flask. The space in the flask up to D contained 4 ounces, therefore the fifth part of the air had been lost. I poured a few ounces of lime water into the flask in order to see whether any aerial acid had also been produced during the combustion, but I did not find any. I made the same experiment with zinc filings, and it proceeded in every way similarly to that just mentioned. I shall demonstrate the constituents of

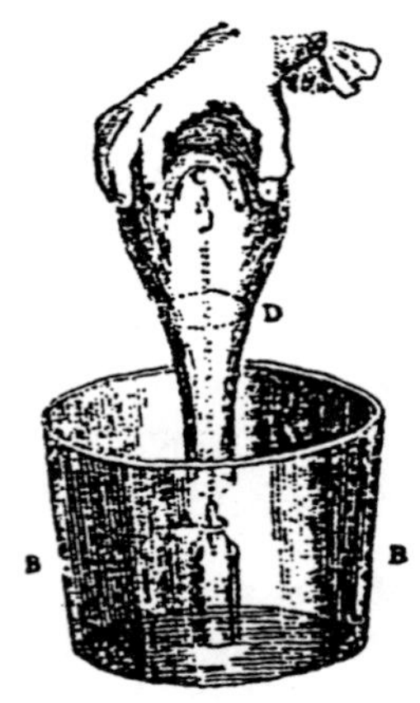

Fig. 1

this inflammable air further on; for, although it seems to follow from these experiments that it is only phlogiston, still other experiments are contrary to this.

We shall now see the behaviour of air towards that kind of fire which gives off, during the combustion, a fluid resembling air. [...]

12.3 E. D. Clarke, *Travels in Various Parts of Europe, Asia and Africa*, Part 3, vol. 11, 1824, pp. 8–10, 130

[...] Afterwards we saw the Chemical Schools in the house of Professor *John Afzelius*, brother of *Adam Afzelius* the botanist, whom we had before visited. He was delivering a lecture, at the time of our arrival, to about twenty or thirty students; but in a voice so low and inaudible, as to be scarcely intelligible, even to those who were his constant hearers. We observed a few among them making notes; but the chief part of the audience seemed to be very inattentive, and to be sitting rather as a matter of form than for any purpose of instruction. Their slovenly dress, and manner, were moreover so unlike that of the students in our *English* Universities, that it was impossible to consider them as gentlemen: they had rather the air and appearance of so many labouring artificers, and might have been mistaken for a company of workmen in a manufactory. Around this chemical lecture-room was arranged the Professor's collection of minerals – perhaps more worthy of notice than any thing else in *Upsala*; for the Chemical Laboratory scarcely merits attention. It was classed according to the methodical distribution of *Cronstedt*, and has been in the possession of the University ever since the middle of the eighteenth century. The celebrated *Bergmann* added considerably to this collection, which may be considered as one of the most complete in Europe; especially in specimens from the *Swedish* mines, which have long produced the most remarkable minerals in the world. One cabinet alone contained three thousand specimens; and the whole series occupied no less a number than forty. It is true, that, in this immense collection, there were many things denoting an earlier period in the history of mineralogy, and which now belong rather to the study of *geology* than of *mineralogy*. One small cabinet contained models of mining apparatus; pumps, furnaces, &c. There is no country that has afforded better proofs of the importance of mineralogical studies to the welfare of a nation, than *Sweden*; but the *Swedes* have not maintained the pre-eminence in *mineralogy* which they so honourably acquired. The *mineralogy* of *Cronstedt* laid the true foundation of the science, by making the chemical composition of minerals the foundation of the species into which they are divided: and whenever an undue regard for the mere external characters of these bodies causes an attention to their chemical constituents to be disregarded, it may be regretted, as an effectual bar to the progress of mineralogical knowledge. [...]

Upon this our second visit to *Stockholm*, we again examined the collection of minerals belonging to the Crown; and were much indebted to the celebrated chemist *Hjelm*, for the readiness he always shewed to gratify our curiosity; allowing us to inspect all the produce of the *Swedish* mines. The refractory nature of some of the richest *iron* ores of this country and of *Lapland* is owing to the presence of several remarkable extraneous bodies; among which may be mentioned *titanium, zircon*, and *phosphate of lime*. We had made a large collection of these ores, and the nature of them is now well ascertained. In the account we gave of our first visit to this collection, a specimen was slightly alluded to, exhibiting a remarkable prismatic configuration, taken from the bottom of a furnace in *Siberia*. How it was brought to *Stockholm* we did not learn. Some of the *Swedish* mineralogists attached more importance to this artificial appearance than we did; considering it as a satisfactory elucidation of the origin of what is commonly called the *basaltic* formation by means of igneous fusion. [. . .]

Chapter Thirteen
Science in Orthodox Europe

13.1 L. Euler, *Letters of Euler to a German Princess,* 1795, trans. Henry Hunter, Letters XVIII, LXXXIX, pp. 79–83, 388–90

LETTER XVIII.

Difficulties attending the System of Emanation.

HOWEVER strange the doctrine of the celebrated *Newton* may appear, that rays proceed from the sun, by a continual emanation, it has, however, been so generally received, that it requires an effort of courage to call it in question. What has chiefly contributed to this, is, no doubt, the high reputation of the great English philosopher, who first discovered the true laws of the motions of the heavenly bodies: and it is this very discovery which led him to the system of emanation.

Descartes, in order to support his theory, was under the necessity of filling the whole space of the heavens with a subtile matter, through which all the celestial bodies move at perfect liberty. But it is well known, that if a body moves in air, it must meet with a certain degree of resistance; from which *Newton* concluded, that, however subtile the matter of the heavens may be supposed, the planets must encounter some resistance in their motions. But, said he, this motion is not subject to any resistance: the immense space of the heavens, therefore, contains no matter. A perfect vacuum, then, universally prevails. This is one of the leading doctrines of the Newtonian philosophy, that the immensity of the universe contains no matter, in the spaces not occupied by the heavenly bodies. This being laid down, there is between the sun and us, or at least from the sun down to the atmosphere of the earth, an absolute vacuum. In truth, the father we ascend, the more subtile we find the air to be; from whence it would apparently follow, that at length the air would be entirely lost....

Having established, then, a perfect vacuum between the heavenly bodies, there remains no other opinion to be adopted, but that of emanation: which obliged *Newton* to maintain, that the sun, and all other luminous bodies, emit rays, which are always particles, infinitely small, of their mass, darted from them with incredible force. It must be such to a very high degree, in order to impress on rays of light that inconceivable velocity with which they come from the sun to us, in the space of eight minutes. But let us see whether this theory be consistent with Newton's leading doctrine, which requires an absolute vacuum in the heavens, that the planets may encounter no manner of resistance to their motions. You must conclude on a moment's reflection, that the space in which the heavenly bodies revolve, instead of remaining a vacuum, must be filled with the rays, not only of the sun, but likewise of all the other stars which are continually passing through it, from every quarter, and in all directions, with incredible rapidity. The heavenly bodies which traverse these spaces, instead of encountering a vacuum, will meet with the matter of luminous rays in a terrible agitation, which must disturb these bodies in their motions, much more than if it were in a state of rest.

Thus *Newton*, apprehensive lest a subtile matter, such as *Descartes* imagined, should disturb the motions of the planets, had recourse to a very strange expedient, and quite contradictory to his own intention, as, on his hypothesis, the planets must be exposed to a derangement infinitely more considerable. I have already submitted to you several other insuperable objections to the system of emanation; and we have now seen that the principal, and indeed the only reason, which could induce *Newton* to adopt it, is so self-contradictory as wholly to overturn it. All these considerations united, leave us no room to hesitate about the rejection of this strange system of the emanation of light, however respectable the authority of the philosopher who invented it.

Newton was, without doubt, one of the greatest geniuses that ever existed. His profound knowledge, and his acute penetration into the most hidden mysteries of nature, will be a just object of admiration to the present, and to every future age. But the errors of this great man should serve to admonish us of the weakness of the human understanding, which, after having soared to the greatest possible heights, is in danger of plunging into manifest contradiction.

If we are liable to weaknesses and inconsistences so humiliating, in our researches into the phenomena of this visible world, which lies open to the examination of our senses, how wretched must we have been, had God left us to ourselves with respect to things invisible, and which concern our eternal salvation? On this important article, a Revelation was absolutely necessary to us; and we ought to avail ourselves of it, with the most profound veneration. When it presents to us things which may appear inconceivable, we have but to reflect on the imperfection of human understanding, which is so apt to be misled, even as to sensible objects. Whenever I hear a pretended Freethinker inveighing against the truths of religion, and even sneering at it with the most arrogant self-sufficiency, I say to myself: poor weak mortal, how inexpressibly more noble and sublime are the subjects which you treat so lightly, than those respecting which the great *Newton* was so grossly mistaken! I could wish your Highness to keep this reflection ever in remembrance: occasions for making it occur but too frequently.

LETTER LXXXIX.

Of the Question respecting the best World possible; and of the Origin of Evil.

YOU know well, that it has been made a question, Whether this world be the best possible? It cannot be doubted, that the world perfectly corresponds to the plan which God proposed to himself, when he created it.

As to bodies, and material productions, their arrangement and structure are such, that certainly they could not have been better. Please to recollect the wonderful structure of the eye, and you will see the necessity of admitting, that the conformation of all its parts is perfectly adapted to fulfil the end in view, that of representing distinctly exterior objects. How much address is necessary to keep up the eye in that state, during the course of a whole life? The juices which compose it must be preserved from corruption; it was necessary to make provision, that they should be constantly renewed, and maintained in a suitable state.

A structure equally marvellous is observable in all the other parts of our bodies, in those of all animals, and even of the vilest insects. [...]

We discover the same perfection in plants: every thing in them concurs to their formation, to their growth, and to the production of their flowers, of their fruits, or of their seeds. What a prodigy to behold a plant, a tree, spring from a small grain, cast into the earth, by the help of the nutritious juices with which the soil supplies it? The productions found in the bowels of the earth are no less wonderful: every part of nature is capable of exhausting our utmost powers of research, without permitting us to penetrate all the wonders of its construction. Nay, we are utterly lost, while we reflect, how every substance, earth, water, air, and fire, concur in the production of all organized bodies; and, finally, how the arrangement of all the heavenly bodies is so admirably contrived, as perfectly to fulfil all these particular destinations.

After having reflected in this manner, it will be difficult for you to believe, that there should have been men who maintained, that the universe was the effect of mere chance, without any design. But there always have been, and there still are, persons of this description; those, however, who have a solid knowledge of nature, and whom fear of the justice of God does not prevent from acknowledging Him, are convinced, with us, that there is a Supreme Being, who created the whole universe, and, from the remarks which I have just been suggesting to you, respecting bodies, every thing has been created in the highest perfection.

13.2 S. P. Krasheninnikov, *The History of Kamtschatka* (trans. James Grieve), from *Opisanie zemli Kamchatki*, 1764, pp. 76–7, 88–9, 120–3, 132–3, 135, 164–5, 176–9, 220–4

The *Kamtschadales* esteem all the burning mountains, and places where hot springs arise, as the habitations of spirits, and approach them with fear; but, as the latter are the most dangerous, they are under the greatest awe of them; and therefore they never willingly discover them to any *Russian*, lest they should be obliged to accompany him near them. It was by chance that I heard of them after I had travelled 100 versts from the place; but this natural phænomenon appeared so curious that I returned to examine it. The people of *Shematchinski* village were obliged to declare the true reason why they had not formerly discovered them, and much against their will were forced to shew me the place, but would not go near it: and when they saw that we lay in the water, drank it, and eat things boiled with it, they expected to see us perish immediately; but when they perceived this did not happen, they told it in the village as an uncommon wonder, and looked upon us as very extraordinary people, since even the devils could not hurt us. [...]

The wild garlick is not only useful in the kitchen, but also in medicine. Both the *Russians* and *Kamtschadales* gather great quantities, which they cut and dry in the sun for their winter provision; at which time boiling it in water they ferment it a little, and use it as an herb soup, which they call *shami*. They esteem the wild garlick so efficacious a remedy against the scurvy, that they think themselves in no danger so soon as it begins to shew itself under the snow: and I have heard an extraordinary account of its virtues from the Cossacks that were employed with captain *Spanberg* in building the sloop Gabriel: they were so ill with the scurvy, that scarce any were able to work, or even to walk, so long as the ground was covered with snow; but as soon as the high lands began to appear green, and the wild garlick to sprout out, the Cossacks fed upon it greedily. Upon their first eating it, they were covered over with scabs in such a manner, that the captain believed they were all infected with the venereal disease. In about a fortnight, these scabs fell off, and they were perfectly recovered of the scurvy. [...]

Some call the sea lions sea horses, because they have manes. In their shape they are like the sea calf; and their necks are bare, excepting a small mane of hard curled hairs: the rest of their body is covered with a chesnut-coloured hair. They have a middle-sized head, short ears, a snout short and drawn up like a pug dog's, great teeth, and webbed feet. They are found most frequently about rocky shores or rocks in the sea, upon which they climb very high, in great numbers. They roar in a strange, frightful manner, much louder than the sea calf; and they are thus far of use to people at sea, that in foggy weather, by their roaring, they warn them of rocks or islands being near, as few rocks or islands in this part of the world are without these animals.

Although in appearance and size this animal seems to be very dangerous, and marches with such a fierce mien that he looks like a true lion, yet is he such a coward, that at the sight of a man he hurries into the water; and when he is surprised asleep, and awakened either by a loud cry or blows with a club, he is in such fear and confusion, that in running away he falls down, all his joints quaking with terror; but, when he finds no possibility of escaping, he will then attack his enemy with the greatest fierceness, shaking his head and roaring very terribly; and then the boldest must seek to save himself from his rage. [...]

Although these animals naturally run from a man, yet it has been observed that they are not always so wild; particularly when their young have scarcely learned to swim. Mr. *Steller* lived six days in a high place amongst whole herds of them, and out of his hut saw several of their actions. The animals lay around him, seeming to observe his fire and what he was employed about; and never ran away, although he even went amongst them, and seized some of their young for his dissections, but remained quite at their ease. [...]

Besides those already described, there are several other sea animals here, the most remarkable of which is the manati, or sea cow. This animal never comes out upon the shore, but always lives in the water; its skin is black and thick, like the bark of an old oak, and so hard that one can scarcely cut it with an ax; its head in proportion to its body is small, and falls off from the neck to the snout, which is so much bent that the mouth seems to lie below; towards the end the snout is white and rough, with white whiskers about nine inches long; it has no teeth, but only two flat white bones, one above, the other below.... The length of the manati is about 28 feet, and its weight about 200 pood. These animals go in droves in calm weather near the mouths of rivers; and though the dams oblige their young always to swim before them, yet the rest of the herd cover them upon all sides, so that they are constantly in the middle of the drove. In the time of flood they come so near the shore, that one may strike them with a club or spear; nay, the author relates that he has even stroked their backs himself with his hand. When they are hurt they swim off to sea, but presently return. [...]

There is such a plenty of manati in *Bering*'s island, that it is sufficient to maintain all the people of *Kamtschatka*. Their flesh, though it takes a long time to boil, tastes well, and is something like beef. The fat of the young resembles pork, and the lean is like veal. [...]

As *Kamtschatka* abounds with lakes and marshes, the swarms of insects in the summer time would make life intolerable there, if it were not for the frequent winds and rains. The maggots are so numerous as to occasion great destruction to their provisions, particularly in the time of preparing their fish, which are sometimes entirely destroyed by them. In the months of *June*, *July*, and *August*, when the weather happens to be fine, the musketoes and small gnats are very troublesome; however the inhabitants do not suffer much from them, as they are at that time, upon account of the fishery, out at sea, where by reason of the cold and wind few of these insects are to be met with. [...]

There are few spiders in *Kamtschatka*; so that the women who are fond of having children, and who have a notion that these insects swallowed render them fruitful and their labour easy, have great trouble to find them. Nothing plagues the natives in their huts so much as the lice and fleas; the women suffer most from the former, by wearing very long, and sometimes false hair. Mr. *Steller* was told, that near the sea is found an insect that resembles a louse, which working itself through the skin into the flesh is never to be cured, unless by cutting the creature intirely out; and that the fishers are very much afraid of them. [...]

Although in outward appearance they resemble the other inhabitants of *Siberia*, yet the *Kamtschadales* differ in this, that their faces are not so long as the other *Siberians'*, their cheeks stand more out, their teeth are thick, their mouth large, their stature middling, and their shoulders broad, particularly those people who inhabit the sea coast.

Their manner of living is slovenly to the last degree; they never wash their hands nor face, nor cut their nails; they eat out of the same dish with the dogs, which they never wash; every thing about them stinks of fish; they never comb their heads, but both men and women plait their hair in two locks, binding the ends with small ropes: when any hair starts out, they sow it with threads to make it lie close; by this means they have such a quantity of lice that they can scrape them off by handfuls, and they are nasty enough even to eat them. Those that have not natural hair sufficient wear false locks, sometimes as much as weigh ten pounds, which makes their heads look like a haycock. [...]

They have filled almost every place in heaven and earth with different spirits, which they both worship and fear more than God: they offer them sacrifices upon every occasion, and some carry little idols about them, or have them placed in their dwellings; but, with regard to God, they not only neglect to worship him; but, in case of troubles and misfortunes, they curse and blaspheme him.

They keep no account of their age, though they can count as far as one hundred; but this is so troublesome to them that without their fingers they do not tell three. It is very diverting to see them reckon more than ten; for having reckoned the fingers of both hands they clasp them together, which signifies ten; then they begin with their toes, and count to twenty; after which they are quite confounded, and cry, *Matcha?* that is, Where shall I take more. They reckon ten months in the year, some of which are longer and some shorter; for they do not divide them by the changes of the moon, but by the order of particular occurrences that happen in those regions. [...]

They do not distinguish the days by any particular appellation, nor form them into weeks or months, nor yet know how many days are in the month or year. They mark their epochs by some remarkable thing or other, such as the arrival of the *Russians*, the great rebellion, or the first expedition to *Kamtschatka*. They have no writings, nor hieroglyphick figures, to preserve the memory of any thing; so that all their knowledge depends upon tradition, which soon becomes uncertain and fabulous in regard to what is long past.

They are ignorant of the causes of eclipses, but when they happen, they carry fire out of their huts, and pray the luminary eclipsed to shine as formerly. They know only three constellations; the Great Bear, the Pleiades, and the three stars in Orion; and give names only to the principal winds. [...]

The burial of the dead, if one can call throwing them to the dogs a burial, is different here from what it is in any other part of the world; for instead of burning or laying the dead bodies in some hole, the *Kamtschadales* bind a strap round the neck of the corps, draw it out of the hut, and deliver it for food to their dogs: for which they give the following reasons; that those who are eaten by dogs will drive with fine dogs in the other world; and that they throw them round near the hut, that evil spirits, whom they imagine to be the occasion of their death, seeing the dead body, may be satisfied with the mischief they have done. [...]

As the *Koreki* and *Kuriles* agree in most of their customs and habits with the *Kamtschadales*, we shall only take notice of those things wherein they differ from them or from one another. The *Koreki*, as is above related, are divided into the rein-deer or wandering *Koreki*; and those that are fixed in one place who live in huts in the earth like the *Kamtschadales*, and in every other respect indeed resemble them; so that whatever remarks we make are to be understood of the wandering *Koreki*, unless otherwise expressed. [...]

There is besides a very great difference in their customs and habits. The wandering *Koreki* are extremely jealous, and sometimes kill their wives upon suspicion only; but when any are caught in adultery, both parties are certainly condemned to death. For this reason the women seem to take pains to make themselves disagreeable; for they never wash their faces or hands, nor comb their hair, and their upper garments are dirty, ragged, and torn, the best being worn underneath. This they are obliged to do on account of the jealousy of their husbands; who say, that a woman has no occasion to adorn herself unless to gain the affections of a stranger, for her husband loves her without that. On the contrary, the fixed *Koreki*, and *Tchukotskoi*, look upon it as the truest mark of friendship, when they entertain a friend, to put him to bed with their wife or daughter; and a refusal of this civility they consider as the greatest affront; and are even capable of murdering a man for such a contempt. This happened to several *Russian* Cossacks before they were acquainted with the customs of the people. The wives of the fixed *Koreki* endeavour to adorn themselves as much as possible, painting their faces, wearing fine cloaths, and using various means to set off their persons. In their huts they sit quite naked, even in the company of strangers.

13.3 Mikhail Lomonosov, *The Appearance of Venus on the Sun, Observed at the St. Petersburg Academy of Sciences on the 16th Day of May in the Year 1761*, Addendum, trans. C. Chant from M. V. Lomonosov, *Izbrannye proizvedeniya*, tom 1: *Estestvennye nauki i filosofiya* (Moscow: Nauka, 1986), pp. 333–6

This rarely encountered phenomenon requires a twofold explanation. First, it is necessary to dispel among uneducated people any unfounded doubts and fears, which are sometimes the cause of general unrest. Not infrequently, heads filled with credulity heed with horror prophecies made on the basis of such celestial phenomena by itinerant beggar-women, who not only have never heard of the word astronomy throughout the whole of their long lives, but can scarcely with their stooping walk glance up at the sky. The stupidity of such ignorant soothsayers and their superstitious listeners should be treated with nothing but contemptuous laughter. Anyone who is disturbed by such dire prognostications deserves his anxiety as a punishment for his own witlessness. But this pertains more to the common people, who have no understanding of the sciences. The peasant ridicules the astronomer as an empty trifler. The astronomer is inwardly amused, realizing how far in his own knowledge he surpasses a man created so like himself.

The second explanation extends to literate people, to readers of Scripture, and devotees of Orthodoxy, whose sacred concerns are laudable in themselves, even if they have sometimes through excess hindered the progress of the higher sciences.

On reading about the extensive atmosphere around the aforementioned planet described here, one of them will say that on account of it, vapours rise, clouds form, rains fall, streams run and collect into rivers, rivers flow into the sea, various plants spring up everywhere, and animals feed on them. And this, like the Copernican system, is against divine law.

Thinking like this led to a similar controversy about the mobility of the Earth. Theologians of the Western Church interpret the words in Joshua, Chapter 10, verse 12, in a precisely literal sense, and on that basis attempt to prove that the Earth is motionless.

But this controversy began not with Christian but with pagan scholars. Of the ancient astronomers (long before the birth of Christ), Nicetas of Syracuse recognized the daily rotation of the Earth about its axis, and Philolaus its annual revolution around the sun. A hundred years later, Aristarchus of Samos gave a clearer account of the solar system. However, the high priests of Hellenic superstition opposed his account, and suppressed the truth for many centuries. The first Cleanthes denounced Aristarchus for daring with his moving-Earth system to displace the great goddess Vesta, sustainer of the entire Earth, and for daring to put into an incessant spin Neptune, Pluto, Ceres, all the nymphs, and the gods of woods and household throughout the entire Earth. And so, the superstition of idolaters held the

Earth of the astronomers in its jaws, preventing it from moving, even though its natural path always coincided with God's command. Meanwhile, the astronomers were obliged to devise ridiculous cycles and epicycles (circles and subsidiary circles), contrary to the mechanics and geometry of the planets' orbits.

A pity that in those times there were no cooks as sharp-witted as the following:

Once, feasting, two astronomers were seated,
And argued 'mongst themselves in language heated.
Earth turning travels round the Sun, did one maintain,
The other that Sun leads the planets in its train.
This one was Ptolemy, the first, Copernicus.
The smiling Cook resolved their quarrel thus:
Knowst thou the course of stars? the host inquired?
Then how to solve this question art inspired?
Copernicus was right, the answer went.
I'll prove it true, although I've never spent
Time on the Sun. What Cook of brains could boast
So few, to turn the Hearth about the Roast?*

In the end, Copernicus revived the solar system, which now bears his name, and demonstrated its marvellous utility in astronomy. Afterwards, Kepler, Newton and other great mathematicians and astronomers raised it to the level of accuracy we now see in predictions of celestial phenomena, a level quite impossible to achieve using a geostatic system.

Although the untold wisdom of divine activity is apparent from the study of all His creatures, in which the physical sciences lead the way, the magnificence and power of His understanding is shown above all by astronomy, which demonstrates order in the behaviour of the heavenly bodies. We have a clearer conception of the Creator, the more closely observations match our predictions; and the better we comprehend our discoveries, the louder we glorify Him.

Holy Scripture should not always be understood literally, but sometimes in a rhetorical sense. St. Basil the Great gives examples of its concord with nature, and in his Homilies of the Hexaemeron clearly shows how the words of the Bible should be interpreted in such instances. He writes in his Homily on the Earth: 'If you hear in the Psalms: "I bear up the pillars of it", take it to mean the power which sustains it' (Homily 1). Discussing God's words and commands on the creation of the World: 'And God said' and so on, he declares as follows: 'Such words are needed for the mind itself to communicate with others' (Homily 2), clearly explaining that divine words do

---

* Rhyming translation was taken from B. N. Menshutkin, *Russia's Lomonosov* (Princeton, NJ: Princeton University Press, 1952), p. 149.

not require mouth, ears or breath for the communication of this goodwill, but that He holds forth by the force of intellect. And in another place (Homily 3), he makes the same point about the clarification of such passages: 'In the curses on Israel, it is said: "And thy heaven above thy head shall be brass". What does this mean? A total drought and an absence of aerial waters.' Interpreting the frequent Biblical references to God's feelings, he writes thus: '"And God saw that it was good": it is not that a certain pleasing aspect of the sea literally presented itself to God. It is not with eyes that the Creator views the beauty of his works; he sees them as they are in his ineffable wisdom.' Is it not enough that here this great and holy man has shown that the interpretation of the sacred books is not only allowed, but indeed necessary, whenever the use of metaphorical expressions leads to an apparent inconsistency with nature?

Truth and faith are two kindred sisters, the daughters of one supreme parent: there can never be any discord between them, unless some slanderer, out of vanity and in an attempt to display his own wisdom, imputes enmity to them. But reasonable and good persons must search for means to explain and avert all so-called conflict between them, following the aforementioned wise teacher of our Orthodox Church. Agreeing with him, St. John of Damascus, the profound theologian and writer of lofty religious verse, wrote in...The Orthodox Faith (Bk. 2, Ch. 6) concerning various opinions on the structure of the world: 'However, whether it is this way or some other, everything exists and is established by God's command.' That is, physical discussions about the structure of the world serve to glorify God, and are not inimical to religion. The same writer declares as follows: 'So there is a heaven of heaven, which is the first heaven a little above the firmament. Thus there are two heavens, for God has also called the firmament heaven. But it is also customary in Holy Scripture to call the air heaven, since it is seen above. It says, "O all ye fowls of the heaven, bless the Lord", meaning "of the air", for the air is not heaven, but a medium for things which fly. Thus there are three heavens, as the divine Apostle says. Even if you wish to accept the seven spheres as seven heavens, there is still nothing inimical to the word of truth.' That is, even if one accepts the ancient Hellenic teachings about the seven heavens, this is not contrary to the Holy Scripture or the words of St. Paul [...].

These great luminaries thus strove to harmonize natural science with religion, reconciling the achievements of the former with certain divinely-inspired thoughts expressed in the Bible within the limits of contemporary astronomical knowledge. If only our present astronomical instruments had been invented then, and numerous observations made by men incomparably surpassing the ancient astronomers in their knowledge of the heavenly bodies, if only thousands of new stars and other phenomena had been discovered, with what flights of spirituality and marvellous eloquence would these holy rhetoricians have preached the wisdom, power and majesty of God!

Some people ask, if there are people like us living on the planets, what is their religion? Is the Gospel preached to them? Are they baptized in the Christian faith? I answer them with a question: In the great southern lands, the shores of which have almost only been observed by navigators, what is the religion of the people living there, and also of the people of other unknown lands, whose appearance, language and customs differ markedly from ours? Who has preached the Gospel to them? If anyone wishes to find out about this, or to convert and baptize them, let him go there in keeping with the words of the Gospel ('Provide neither gold, nor silver, nor brass in your purses, nor scrip for your journey, neither two coats, neither shoes, nor yet staves'). And as he finishes his sermon, then let him go afterwards to Venus for the same purpose. If only his work will not be in vain. Perhaps the people there have not sinned in Adam, and none in consequence have any need of this. 'There are many roads to salvation. There are many dwellings in heaven.'

For all this, the Christian faith remains indisputable. It cannot be in opposition to the divine Creation, nor the divine Creation to it, even though opposition is promoted by those who have not studied the divine Creation in sufficient depth.

The Creator has given the human race two books. In one, he displayed his greatness, in the other his will. The first is this visible world, created by Him so that man, observing the immensity, beauty and symmetry of his structures, should recognize divine omnipotence, within the limits of understanding granted him. The second book is the Holy Scripture. In it is shown the Creator's goodwill towards our salvation. The great church teachers are the interpreters and expounders of these divinely-inspired books of the prophets and apostles. The physicists, mathematicians, astronomers and other expounders of the divine acts influential in nature are to the book of the constitution of this visible world, what the prophets, apostles and church teachers are to the other book. The mathematician reasons incorrectly, if he wishes to measure the divine will with a pair of compasses. So does the teacher of theology, if he thinks that one can learn astronomy or chemistry from the psalter.

The interpreters and preachers of the Holy Scripture show the path to virtue, depict the rewards of the righteous, the punishment of lawbreakers and the happiness of the life which accords with the divine will. Astronomers reveal the temple of divine power and majesty, and search out means for our temporal welfare, uniting it with reverence and gratitude toward the Supreme Being. Both in general convince us not only of God's existence, but also of His untold kindness to us. It is a sin to sow the seeds of dissension between them!

The extent to which the discussion and investigation of natural phenomena strengthens faith is shown in passages not only from the Hellenic poets, but also from the great teachers of the early Christian Church. [...]

Little remains, except briefly to repeat that knowledge of nature, whatever its name, is not contrary to Christian law; and whoever endeavours to investigate nature, and knows and venerates God, will agree with Basil the Great, with whose words this piece ends (Homily 6, on the existence of heavenly bodies): 'If we are to learn from this, we shall know ourselves, we shall know God, we shall adore our Creator, we shall work for our Master, we shall glorify our Father, we shall love our Provider, we shall honour our Benefactor, and we shall not cease to worship the Prince of our present and future life.'

Chapter Fourteen
Establishing Science in Eighteenth-Century Europe

14.1 Robert Jameson, 'On the Supposed Existence of Mechanical Deposits and Petrefactions in the Primitive Mountains, and an Account of Petrefactions which have been discovered in the newest Flötz Trapp Formation', *Journal of Natural Philosophy, Chemistry and the Arts,* [Nicholson's Journal] 3, 1802, pp. 13–21

On the supposed Existence of Mechanical Deposits and Petrefactions in the Primitive Mountains, and an Account of Petrefactions which have been discovered in the newest Flötz Trapp Formation. By Mr. ROBERT JAMESON. Communicated by the Author.

The primitive rocks precipitated from a great depth of water which covered the globe.

OUR globe, according to the Wernerian geognosia, even during the decomposition of the newer primitive strata, appears to have been covered to a great height with water, as is evinced by the want of all mechanical deposit.* After the precipitation of these great rock formations, the level of the water became so low, as to allow it to act mechanically upon the subjacent rocks; this occasioned the first mechanical deposition, which discovers itself in the transition rocks (Ubergangsgebürge). Nearly at the same time organization commenced, as it is in the transition rocks we find the first traces of organic remains: these are generally zoophytes and sea plants, a fact which goes deep not only into geology, but natural history.

The transition rocks when the water was lower: at which time organization began.

---

* The exception to this in the sienite formation I explained in a former paper.

Instances of mechanical deposits in primary strata: offered by Professor Playfair.

Professor Playfair, in his illustrations of the Huttonian theory, mentions several instances of mechanical deposits and petrefactions which have been discovered among the primary strata, and from these he concludes, that no such series of strata as the transition exist. I shall now examine the statements he has given; and first respecting the occurrence of petrefactions in primary mountains.

Shells in primitive limestone, in the vicinity of Plymouth, &c.

At page 164 he observes, 'Another spot, affording instances of shells in primitive limestone, is in Devonshire, on the sea shore, on the east side of Plymouth dock, opposite to Stonehouse, I found a specimen of shistose micaceous limestone, containing a shell of the bivalve kind; it was struck off from the solid rock, and cannot possibly be considered as an adventitious fossil. Now, no rocks can be more decidedly primary than those about Plymouth; they consist of calcareous strata, in the form either of marble or micaceous limestone, alternating with shistus of the same kind, which prevails through Cornwall to the west, and extends eastward into Dartmoor, and on the sea coast as far as Berry Head. These all intersect the horizontal plane in a line from east to west nearly; they are very erect, those at Plymouth being elevated to the north.' That petrefactions exist in the limestone at Plymouth is evident; but that these strata are primitive, still remain to be proved. The character given of the limestone does not exclude it from the transition strata; but of the shistus we cannot judge, as neither its oryctognostical, or geognostical characters are given. The other instances which are alluded to, are liable to the same objection. I cannot therefore agree with Professor Playfair in believing 'Though, therefore, the remains of marine animals are not frequent among the primary rocks, they are not excluded from them; and hence the existence of shell fish and zoophytes, is clearly proved to be anterior to the formation even of those parts of the present land which are justly accounted the most ancient.'

But it is not proved that those strata are primitive.

Whence the conclusion of the Professor is not agreed to.

The position that vegetable matters are found in primitive strata need not be examined.

Professor Playfair agrees with Dr. Hutton in affirming, that vegetable matters occur in the primitive strata: I do not find it necessary to enter into an examination of what they have said upon this subject, as they have evidently confounded a geognostical with an oryctognostical investigation.

I shall now examine the proofs which Professor Playfair has brought to establish the existence of mechanical deposits in primitive strata.

Prof. Playfair's proofs of mechanical deposits in primitive strata examined.

The first we meet with is from Saussure. Professor Playfair remarks, 'St. Gothard is a central point, in one of the greatest tracts of primary mountains on the face of the earth, yet arenaceous strata are found in its vicinity. Between Ayrolo and the Hospice of St. Gothard, Saussure found a rock, composed of an arenaceous or granular paste, including in it hornblende and garnets. He is somewhat unwilling to give the name gres to this stone, which Mr. Besson has done; but he nevertheless describes it as having a granulated structure.'

Mount St. Gothard, a central primitive mountain, has arenaceous strata in its vicinity.

The rock of Ayrolo is primitive, and is either gneiss or mica slate. Garnets are seldom found in gneiss, but are characteristic for mica slate; the geognost, therefore, would not hesitate to consider the rock here mentioned as belonging to mica slate.

Observation. The rock here mentioned was gneiss or mica slate.

Professor Playfair continues, 'Among the most indurated rocks that compose the mountains of this island, many are arenaceous. Thus, on the western coast of Scotland, the great body of high and rugged mountains on the shores of Arasaig, &c. from Ardnamurchan to Glen-elg, consists, in a great measure, of a granitic sandstone, in vertical beds. This stone sometimes occupies great tracts; at other times, it is alternated with the micaceous, and other varieties of primary shistus; it occurs, likewise, in several of the islands, and is a fossil which we hardly find described or named by writers on mineralogy.'

Instances by P. Playfair of indurated arenaceous granite.

This granitic sandstone of Glen-elg is most certainly gneiss,* and a variety which is not uncommon; and I may venture to say, that the strata in Arasaig, &c. are of the same nature.

Obs. This granitic sandstone is gneiss.

Professor Playfair concludes with stating the following examples, as a further confirmation of his opinion – 'Much also of a highly indurated, but granulated quartz, is found in several places in Scotland, in beds of strata, alternated with the common shistus of the mountain. Remarkable instances of this may be seen on the north side of the ferry of Balachulish, and again on the sea shore at Cullen. At the latter, the strata are remarkably regular, alternating with different species of shistus. At

Other examples of Pr. Playfair. Granulated quartz in beds alternating with the shistus.

* Mineralogy of the Scottish isles, vol. ii, p. 160.

the former, the quartz is so pure, that the stone has been mistaken for marble.

'These examples are perhaps sufficient; but I must add, that in the micaceous and talcose schisti themselves, thin layers of sand are often found interposed between the layers of mica or talc. I have seen a specimen from the summit of one of the highest of the Grampian mountains, where the thin plates, of a talcy or asbestine substance, are separated by layers of a very fine quartzy sand, not much consolidated. The mountain from which it was brought, consists of vertical strata, much intersected by quartz veins. It is impossible to doubt, in this instance, that the thin plates of the one substance, and the small grains of the other, were deposited together at the bottom of the sea, and that they were alike produced from the degradation of rocks more ancient than any which now exist.'

Whence he concludes that they were deposited (mechanically) at the same time.

I am surprised Professor Playfair should adduce granular quartz as a proof of mechanical deposition, as it has no more claim to such a character than granite. The following observations will render this evident; granular quartz differs from mica slate, in the absence of the slaty fracture, and mica; we have, however, a series from the most complete granular quartz to the most perfect mica slate. Again, mica slate, which differs from gneiss, in wanting felspar, is to be observed in all the intermediate stages until it passes into complete gneiss; the gneiss, which is principally distinguished from granite by its slaty fracture, gradually loses this fracture, and at length is not to be distinguished from granite. Thus we have a complete gradation from the purest granular quartz to granite, and not only in hand specimens, but in the mountains themselves. It therefore follows, that if granular quartz is a mechanical deposit, so is granite, a position which I believe the Huttonian system would not allow.

That the granular quartz should occur sometimes of a very loose texture; nay, even as Professor Playfair remarks sandy, is not surprising, for in granite and basalt we have similar appearances. Even in veins, which to use Professor Playfair's own words, bear all the marks of complete fusion, layers of granular quartz, from the most compact to the looseness of sand have been observed. The great sand veins in the Hartz afford remarkable instances of this; also, as I have more lately

Obs. Granular quartz is not a mechanical, but a chemical deposition. For this differs from mica slate in the absence of slaty fracture; but the series between each are gradual. Mica slate differs from gneiss in wanting felspar; but here also the transition is equally gradual. And gneiss differs from granite by its slaty fracture, which it as gradually loses. The gradation from granular quartz to granite being perfect, the one is no more a mechanical deposit than the other. Loose texture is no proof of mechanical deposition; and layers of sand are found in veins which in the volcanic theory are said to bear every mark of complete fusion. General observations.

discovered, the lead veins at Wanlock-head in Lancashire.

Account of Organic Remains which have been discovered in the newest Flötz Trapp Formation.

Many different accounts had been given of the geognostic relations of the rocks of this formation, before they engaged the attention of Werner, the great founder of *true geognosia*. After having made the remarkable discovery upon the hill of Scheibenberg, he extended his inquiries to all the basalt hills in Germany, and found in every quarter corresponding appearances. This confirmed more completely the conclusions he had then drawn, and entirely overthrew the volcanic system. He however did not stop here; his after observations disclosed a connection among these appearances, which at first he was not probably aware of; they placed the Neptunian system beyond the reach of attack, and completely annihilated a host of hypothesis. He proved,

1. That this is the newest of all the great rock formations, of which the crust of the earth is composed.
2. That all the apparently unconnected hills and masses of this formation, have formerly stood in connection with each other.
3. That it exists in all quarters of the globe.

Results. 1. That this is the newest of the great rocks. 2. That all the now unconnected masses of this formation were once connected. 3. That it exists everywhere. 4. That it formerly covered the whole earth, and must contain organic remains.

These facts lead him to the great discovery, that this formation at one time, extended as a cover around the whole earth. From these observations it is evident, that we may expect to meet with vestiges of many of the organic and unorganic matter which at that time existed upon the crust of the earth, in the rocks of this formation.

Accordingly, the investigations of geognosts have discovered organic remains in every rock of this formation; the following are given as instances:

Organic Remains in Greenstone, Basalt, Wacke, and Trapp Buccin.

QUADRUPEDS

1. Werner in his geognostic lectures, informs us that the wacken of Kalten-nordheim is sometimes found to contain deers horns.

Organic remains in greenstone, basalt, &c. found; viz. of quadrupeds.

2. The Abbé Fortis discovered the head of an unknown animal in a soft wacken, in the valley of Ronca in the Veronese. *Vide Beschrebbung des Thales Ronca, s. 96.*
3. Saussure observed bones of quadrupeds in the wacken of the catacombs of Rome.

Lettre a Mr. le Chev. Hamilton, J. de Physicque. Tom. VII.

SHELLS AND ZOOPHYTES

Of shells and zoophytes.

1. Dr. Richardson found shells in rocks of the Trapp formation at Ballycastle in Ireland – *Kirwan's Geol. Essays.*
2. Mr. Von Buch informs us, that in the county of Landeck, he observed a bed of wacken, which contained besides pebbles of chalcedony, turbinates in a state of complete preservation. *Versuch einer mineralogisten Veschreibung von Landeck, von Leopold von Buch, s. 35.*
3. Abbé Fortis observed numerous petrefactions of shells in Wacken, and Trapp Breccia, also a few in Basalt, in the formation of the valley of Ronca.
4. Berolding found a cornu ammonis, which still retained its mother of pearl lustre, in the basalt of Torez. In the basalt of Thurgau, near the Boden lake, he observed gryphites, ammonites, and glossopetræ, *Chem. Annal.* 1794, p. 103.
5. In the Wernerian collection of petrefactions I saw specimens of greenstone, containing petrefactions of shells.

Of vegetables.

VEGETABLE REMAINS

1. Werner observed great trees, with branches, leaves, and fruit, in the wacken, at Joachimsthal.
2. Friesleben describes the impression of a plant in the Kawsower Berg, near to Podsedlitz.
3. In the islands of Barra and Skye I observed pieces of wood in Trapp Breccia, ... Mineralogy of the Scottish isles, vol. ii, p. 58–75.

Organic remains in slaty clay, limestone, and sandstone, of this formation. Shells.

Organic remains found in slaty clay, limestone, and sandstone, which belongs to the Flötz Trapp formation.

SHELLS

1. Abbé Fortis, in his account of the valley of Ronca, informs us, that limestone and slaty clay often

alternate with basalt, wacken, and trapp breccia. The limestone and slaty clay contains numerous petrefactions of shells, which are of the same kinds with those found in the basalt, wacken, and trapp breccia.

2. In the island of Eigg, where there appears to be a similar formation with that of Ronca, I observed that the limestone, slaty clay, and sandstone, contained numerous petrefactions of shells. Mineralogy of the Scottish isles, vol. ii.

VEGETABLE REMAINS

Of vegetables.

1. At the northern extremity of the island of Skye, where basalt alternates with limestone and slaty clay, I observed pieces of carbonated wood in the limestone. Mineralogy of the Scottish isles, vol. ii, p. 80.*
2. In the flinty sandstone, which usually accompanies this formation, I have observed branches of shrubs; vegetable matters that occur inveloped in rocks are generally, either carbonated or bituminated, here however they are not altered.

R. JAMESON.

Sheriff Bræ, Leith.

POSTSCRIPT

Dr. Hutton's statement of facts to shew the mechanical origin of primitive countries from the existence of organic remains.

Since writing the enclosed, I have examined one of those appearances, which are considered by Dr. Hutton and Professor Playfair, as demonstrating the existence of petrefactions in primitive mountains. Dr. Hutton, at p. 334 of his Theory of the Earth, remarks, 'I have already observed, that one single example of a shell, or of its print, in a schistus, or in a stone stratified among those vertical or erected masses, suffices to prove the origin of these bodies to have been, what I had maintained them to be, water formed strata created from the

* Mr. Kirwan, in the first volume of his System of Mineralogy, has given us an excellent account of the different opinions respecting the formation of basalt; and has adduced many arguments that shew the fallacy of the Volcanic and Plutonic hypothesis.

bottom of the sea, like every other consolidated stratum of the earth. But now, I think, I may affirm that there is not, or rarely, any considerable extent of country of the primary kind, in which some mark of this origin will not be found, upon careful examination; and now I will give my reason for this assertion. I have been examining the south alpine country of Scotland occasionally, for forty years back, and I could not find any mark of an organized body in the schistus of those mountains. It is true, that I knew of only one place where limestone is found among the strata: this is upon Tweedside near the Crook. This quarry I had carefully examined long ago, but could find no mark of any organized body in it. I suppose they are now working some other of the vertical strata near to those which I had examined: for, in the summer of 1792, I received a letter from Sir James Hall, which I shall now transcribe. It is dated Moffat, June 2, 1792.

Sir James Hall's account of organic remains in a limestone stated to be primitive.

'As I was riding yesterday between Noble House and the Crook, on the road to this place, I fell in with a quarry of alpine limestone; it consists of four or five strata, about three feet thick, one of them single, and the rest contiguous; they all stand between the strata of slate and schist, that are at that place nearly vertical. In the neighbourhood, a slate quarry is worked of pure blue slate; several of the strata of slate near the limestone, are filled with fragments of limestone scattered about like the fragments of schist in the sandstone, in the neighbourhood of the junction on our coast. Among the masses of limestone lately broken off for use, and having the fracture fresh, I found the forms of cockles quite distinct, and in great abundance. I send you three pieces of this kind,' &c.

It may perhaps be alleged, that those mountains of Cumberland and Tweedale are not the primary mountains, but composed of the secondary schistus, which is every where known to contain these objects belonging to the former earth. Naturalists who have not an opportunity of convincing themselves by their proper examination, must judge with regard to that geological fact by the description of others. Now it is most fortunate for natural history, that it has been in this range of mountains that we have discovered those marks of a marine origin: for, I shall afterwards have occasion to give the clearest light into this subject, from observations made in other parts of those same mountains of schist, by which it will be proved that they are primary strata; and thus no manner of doubt will remain in the minds of naturalists, who might otherwise suspect that we were deceiving ourselves, by mistaking the secondary for the primitive schistus.

Dr. Hutton's account of the mountains in the south of Scotland is confused and unscientifical, and hardly comes within the pale of *true geognosia*. It is not my intention, at present, to enter into an examination of his observations; the object of this postscript is to shew, that the limestone between Noble House and the Crook Inn does not belong to the primitive mountains. The beds of limestone mentioned by Sir James Hall, I observed lying between strata of transition slate, and this slate alternating with strata of grey wacke; consequently the whole belongs to the transition class of rocks.

Remark that the limestone is not primitive,

The limestone has a blueish grey colour, fracture is foliated, the distinct concretions are from coarse to fine grained, and it is hardly translucid on the edges. It is often traversed by veins of calcareous spar, and sometimes it contains thin beds of flinty slate (Kiesel Schiefer of Werner). The transition slate has a blueish or smoke grey colour, has generally less lustre than the primitive slate, and contains much [interspersed] mica. I observed it in all the stages from nearly pure slate to grey wacke. The grey wacke is composed of fragments of transition slate, flinty slate, and quartz, connected by a basis of transition slate. It is frequently traversed by veins of quartz, and is to be observed where the fragments are hardly distinguishable from their size; it has much the appearance of a breccia. I shall take another opportunity of sending you drawings of the different kinds of petrefactions that occur in the limestone.

for it lies between strata of transition slate; alternating with strata of grey wacke. Description of the several rocks.

Chapter Fifteen
The Chemical Revolution

15.1 Stephen Hales, *Vegetable Staticks*, 1727 (London: Science Book Guild, 1961), pp. 89–95

The excellent Mr. *Boyle* made many Experiments on the Air, and among other discoveries, found that a good quantity of Air was producible from Vegetables, by putting Grapes, Plums, Gooseberries, Cherries, Pease, and several other sorts of fruits and grains into exhausted and unexhausted receivers, where they continued for several days emitting great quantities of Air.

Being desirous to make some further researches into this matter, and to find what proportion of this Air I could obtain out of the different substances in which it was lodged and incorporated, I made the following chymio-statical Experiments: For, as whatever advance has here been made in the knowledge of the nature of Vegetables, has been owing to statical Experiments, so since nature, in all her operations, acts conformably to those mechanick laws, which were established at her first institution; it is therefore reasonable to conclude, that the likeliest way to enquire, by chymical operations, into the nature of a fluid, too fine to be the object of our sight, must be by finding out some means to estimate what influence the usual methods of analysing the animal, vegetable, and mineral kingdoms, has on that subtile fluid; and this I effected by affixing to retorts and boltheads hydrostatical gages, in the following manner, *viz.*

In order to make an estimate of the quantity of Air, which arose from any body by distillation or fusion, I first put the matter which I intended to distill into the small retort *r* (Fig. 33) and then at *a* cemented fast to it the glass vessel *ab*, which was very capacious at *b*, with a hole in the bottom. I bound bladder over the cement which was made of tobacco-pipe clay and bean flower, well mixed with some hair, tying over all four small sticks, which served as splinters to strengthen the joynt; sometimes, instead of the glass vessel *ab*, I made use of a large bolthead, which had a round hole cut, with a red hot iron ring at the bottom of it; through which hole was put one

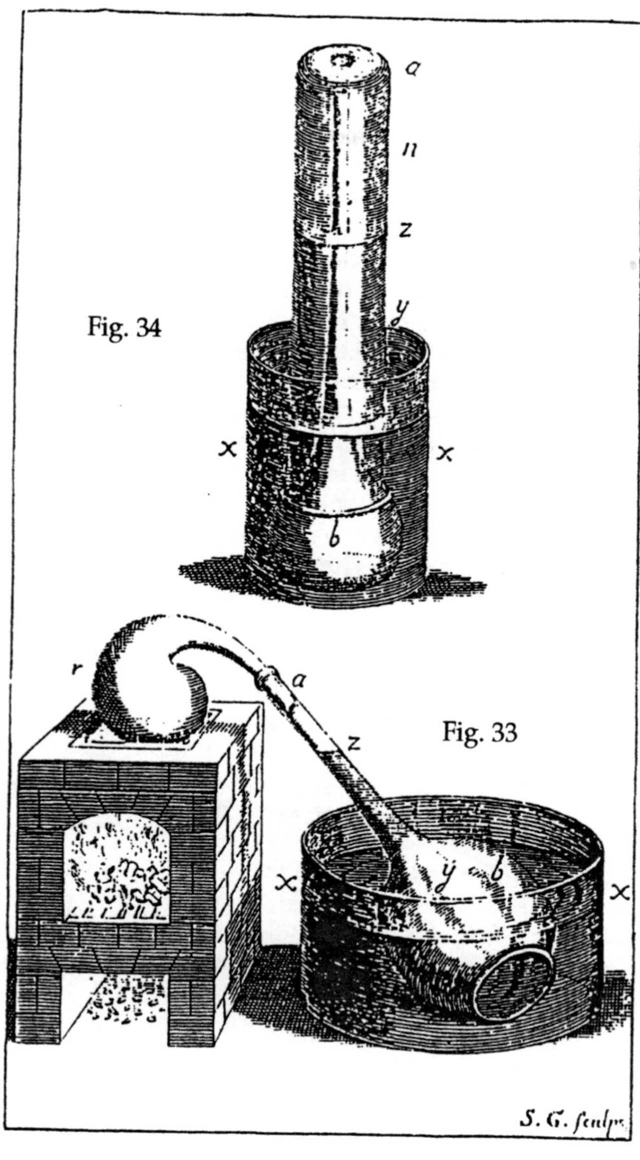

Fig. 34

Fig. 33

leg of an inverted syphon, which reached up as far as *z*. Matters being thus prepared, holding the retort uppermost, I immersed the bolthead into a large vessel of water, to *a* the top of the bolthead; as the water rushed in at the bottom of the bolthead, the Air was driven out thro' the syphon: When the bolthead was full of water to *z*, then I closed the outward orifice of the syphon with the end of my finger, and at the same time drew the other leg of it out of the bolthead, by which means the water continued up to *z*, and could not subside. Then I placed under the bolthead, while it was in the water, the vessel *xx*, which done, I lifted the vessel *xx* with the bolthead in it out of the water, and tyed a waxed thread at *z* to mark the height of the water: And then approached the retort gradually to the fire, taking care to screen the whole bolthead from the heat of the fire.

The descent of the water in the bolthead shewed the sums of the expansion of the Air, and of the matter which was distilling; The expansion of the Air alone, when the lower part of the retort was beginning to be red hot, was at a medium, nearly equal to the capacity of the retorts, so that it then took up a double space; and in a white and almost melting heat, the Air took up a tripple space or something more: for which reason the least retorts are best for these Experiments. The expansion of the distilling bodies was sometimes very little, and sometimes many times greater than that of the Air in the retort, according to their different natures.

When the matter was sufficiently distilled, the retort *&c.* was gradually removed from the fire, and when cool enough, was carried into another room, where there was no fire. When all was thoroughly cold, either the following day, or sometimes 3 or 4 days after, I marked the surface of the water *y*, where it then stood; if the surface of the water was below *z*, then the empty space between *y* and *z* shewed how much Air was generated, or raised from a fix'd to an elastick state, by the action of the fire in distillation: But if *y* the surface of the water was above *z*, the space between *z* and *y*, which was filled with water, shewed the quantity of Air which had been absorbed in the operation, *i.e.* was changed from a repelling elastick to a fix'd state, by the strong attraction of other particles, which I therefore call absorbing. [. . .]

I made use of the following means to measure the great quantities of Air, which were either raised and generated, or absorbed by the fermentation arising from the mixture of variety of solid and fluid substances, whereby I could easily estimate the surprising effects of fermentation on the air, *viz.*

I put into the bolthead *b* (Fig. 34) the ingredients, and then run the long neck of the bolthead into the deep cylindrical glass *ay*, and inclined the inverted glass *ay*, and bolthead almost horizontally in a large vessel of water, that the water might run into the glass *ay*; when it was almost up to *a* the top of the bolthead, I then immersed the bottom of the bolthead, and lower part *y* of the cylindrical glass under water, raising at the same time the end *a* uppermost. Then before I took them out of the water, I set the bolthead and lower part of the cylindrical glass *ay* into the earthen vessel *xx* full of water, and having lifted all out of the great vessel of water, I marked the surface *z* of the water in the glass *ay*.

If the ingredients in the bolthead, upon fermenting generated Air, then the water would fall from *z* to *y*, and the empty space *zy* was equal to the bulk of the quantity of Air generated: But if the ingredients upon fermentation did absorb or fix the active particles of Air, then the surface of the water would ascend from *z* to *n*, and the space *zn*, which was filled with water, was equal to the bulk of Air, which was absorbed by the ingredients, or by the fume arising from them: When the quantities of Air, either generated or absorbed, were very great, then I made use of large chymical receivers instead of the glass *ay*: But if these quantities were very small, then instead of the bolthead and deep cylindrical glass *ay*, I made use of a small cylindrical glass, or a common beer glass inverted, and placed under it a Viol or Jelly glass, taking care that the water did not come at the ingredients in them, which was easily prevented by drawing the water up under the inverted glass to what height I pleased by means of a syphon; [...]

The illustrious Sir *Isaac Newton* (query 31st of his Opticks) observes, that 'true permanent Air arises by fermentation or heat, from those bodies which the chymists call fixed, whose particles adhere by a strong attraction, and are not therefore separated and rarified without fermentation. Those particles receding from one another with the greatest repulsive force, and being most difficultly brought together, which upon contact were most strongly united. And query 30. dense bodies by fermentation rarify into several sorts of Air; and this Air by fermentation, and sometimes without it, returns into dense bodies.' Of the truth of which we have evident proof from many of the following Experiments, *viz.*

That I might be well assured that no part of the new Air which was produced in distillation of bodies, arose either from the greatly heated Air in the retorts, or from the substance of the heated retorts, I first gave a red hot heat both to an empty glass retort and also to an iron retort made of a musket barrel; when all was cold, I found the Air took up no more room than before it was heated: whence I was assured, that no Air arose, either from the substance of the retorts, or from the heated Air.

15.2 Joseph Priestley, *Experiments and Observations on Different Kinds of Air*, 2 vols, vol. 2, 1775 (Edinburgh: Alembic Club Reprint no. 7, 1947), pp. 14–19

Till this 1st of March, 1775, I had so little suspicion of the air from mercurius calcinatus, &c. being wholesome, that I had not even thought of applying to it the test of nitrous air; but thinking (as my reader must imagine I frequently must have done) on the candle burning in it after long agitation in water, it occurred to me at last to make the experiment; and putting one measure of nitrous air to two measures of this air, I found, not only that it was diminished, but that it was diminished quite as much as common air, and that the redness of the mixture was likewise equal to that of a similar mixture of nitrous and common air.

After this I had no doubt but that the air from mercurius calcinatus was fit for respiration, and that it had all the other properties of genuine common air. But I did not take notice of what I might have observed, if I had not been so fully possessed by the notion of there being no air better than common air, that the redness was really deeper, and the diminution something greater than common air would have admitted.

Moreover, this advance in the way of truth, in reality, threw me back into error, making me give up the hypothesis I had first formed, viz. that the mercurius calcinatus had extracted spirit of nitre from the air; for I now concluded, that all the constituent parts of the air were equally, and in their proper proportion, imbibed in the preparation of this substance, and also in the process of making red lead. For at the same time that I made the above mentioned experiment on the air from mercurius calcinatus, I likewise observed that the air which I had extracted from red lead, after the fixed air was washed out of it, was of the same nature, being diminished by nitrous air like common air: but, at the same time, I was puzzled to find that air from the red precipitate was diminished in the same manner, though the process for making this substance is quite different from that of making the two others. But to this circumstance I happened not to give much attention.

I wish my reader be not quite tired with the frequent repetition of the word *surprize*, and others of similar import; but I must go on in that style a little longer. For the next day I was more surprized than ever I had been before, with finding that, after the above-mentioned mixture of nitrous air and the air from mercurius calcinatus, had stood all night, (in which time the whole diminution must have taken place; and, consequently, had it been common air, it must have been made perfectly noxious, and intirely unfit for respiration or inflammation) a candle burned in it, and even better than in common air.

I cannot, at this distance of time, recollect what it was that I had in view in making this experiment; but I know I had no expectation of the real issue of it. Having acquired a considerable degree of readiness in making experiments of this kind, a very slight and evanescent motive would be sufficient to induce me to do it. If, however, I had not happened, for some other purpose, to have had a lighted candle before me, I should probably never have made the trial; and the whole train of my future experiments relating to this kind of air might have been prevented.

Still, however, having no conception of the real cause of this phenomenon, I considered it as something very extraordinary; but as a property that was peculiar to air that was extracted from these substances, and *adventitious*; and I always spoke of the air to my acquaintance as being substantially the same thing with common air. [. . .]

On the 8th of this month I procured a mouse, and put it into a glass vessel, containing two ounce-measures of the air from mercurius calcinatus. Had it been common air, a full-grown mouse, as this was, would have lived in it about a quarter of an hour. In this air, however, my mouse lived a full half hour; and though it was taken out seemingly dead, it appeared to have been only exceedingly chilled; for, upon being held to the fire, it presently revived, and appeared not to have received any harm from the experiment.

By this I was confirmed in my conclusion, that the air extracted from mercurius calcinatus, &c. was, *at least, as good* as common air; but I did not certainly conclude that it was any *better*; because, though one mouse would live only a quarter of an hour in a given quantity of air, I knew it was not impossible but that another mouse might have lived in it half an hour; so little accuracy is there in this method of ascertaining the goodness of air: and indeed I have never had recourse to it for my own satisfaction, since the discovery of that most ready, accurate, and elegant test that nitrous air furnishes. But in this case I had a view to publishing the most generally-satisfactory account of my experiments that the nature of the thing would admit of.

This experiment with the mouse, when I had reflected upon it some time, gave me so much suspicion that the air into which I had put it was better than common air, that I was induced, the day after, to apply the test of nitrous air to a small part of that very quantity of air which the mouse had breathed so long; so that, had it been common air, I was satisfied it must have been very nearly, if not altogether, as noxious as possible, so as not to be affected by nitrous air; when, to my surprize again, I found that though it had been breathed so long, it was still better than common air. For after mixing it with nitrous air, in the usual proportion of two to one, it was diminished in the proportion of $4\frac{1}{2}$ to $3\frac{1}{2}$; that is, the nitrous air had made it two ninths less than before, and this in a very short space of time; whereas I had never found that, in the longest time, any common air was reduced more than one fifth of its bulk by any proportion of nitrous air, nor more than one fourth by any phlogistic process whatever. Thinking of this extraordinary fact upon my pillow; the next morning I put another measure of nitrous air to the same mixture, and, to my utter astonishment, found that it was farther diminished to almost one half of its original quantity. I then put a third measure to it; but this did not diminish it any farther: but, however, left it one measure less than it was even after the mouse had been taken out of it.

Being now fully satisfied that this air, even after the mouse had breathed it half an hour, was much better than common air; and having a quantity of it still left, sufficient for the experiment, viz. an ounce-measure and a half, I put the mouse into it; when I observed that it seemed to feel no shock upon being put into it, evident signs of which would have been visible, if the air had not been very wholesome; but that it remained perfectly at its ease another full half hour, when I took it out quite lively and vigorous. Measuring the air the next day, I found it to be reduced from $1\frac{1}{2}$ to $\frac{2}{3}$ of an ounce-measure. And after this, if I remember well (for in my *register* of the day I only find it noted, that it was *considerably diminished* by nitrous air) it was nearly as good as common air. It was evident, indeed, from the mouse having been taken out quite vigorous, that the air could not have been rendered very noxious.

For my farther satisfaction I procured another mouse, and putting it into less than two ounce-measures of air extracted from mercurius calcinatus and air from red precipitate (which, having found them to be of the same quantity, I had mixed together) it lived three quarters of an hour. But not having had the precaution to set the vessel in a warm place, I suspect that the mouse died of cold. However, as it had lived three times as long as it could probably have lived in the same quantity of common air, and I did not expect much accuracy from this kind of test, I did not think it necessary to make any more experiments with mice.

15.3 Joseph Priestley, *Autobiography of Joseph Priestley: Memoirs Written by Himself*, 1806, intro. J. Lindsay (Bath: Adams and Dart, 1970), pp. 128–33

A continuation of the Memoirs, written at Northumberland, in America, in the beginning of the year 1795.

When I wrote the preceeding part of these Memoirs, I was happy, as must have appeared in the course of them, in the prospect of spending the remainder of my life at Birmingham, where I had every advantage for pursuing my studies, both philosophical and theological; but it pleased the sovereign disposer of all things to appoint for me other removals, and the manner in which they were brought about, were more painful to me than the removals themselves. I am far, however, from questioning the wisdom or the goodness of the appointments respecting myself or others.

To resume the account of my pursuits, where the former part of the Memoirs left it, I must observe that, in the prosecution of my *experiments*, I was led to maintain the doctrine of phlogiston against Mr. Lavoisier, and other chemists in France, whose opinions were adopted not only by almost all the philosophers of that country, but by those in England and Scotland also. My friends, however, of the lunar society, were never satisfied with the anti-phlogistic doctrine. My experiments and observations on this subject, were published in various papers in the 'Philosophical Transactions'. At Birmingham I also published a new edition of my publications on the subject of *air*, and others connected with it, reducing the six volumes to three, which, with his consent, I dedicated to the Prince of Wales.

In theology, I continued my 'Defences of Unitarianism', until it appeared to myself and my friends, that my antagonists produced nothing to which it was of any consequence to reply. But I did not, as I had proposed, publish any address to the bishops, or to the legislature, on the subject. The former I wrote but did not publish. I left it, however, in the hands of Mr. Belsham, when I came to America, that he might dispose of it as he should think proper.

The pains that I took to ascertain the state of early opinions concerning Jesus Christ, and the great misapprehensions I perceived in all the ecclesiastical historians, led me to undertake a 'General History of the Christian Church to the Fall of the Western Empire', which accordingly I wrote in two volumes octavo, and dedicated to Mr. Shore. This work I mean to continue.

At Birmingham I wrote the 'Second Part' of my 'Letters to a Philosophical Unbeliever', and dedicated the whole to Mr. Tayleur, of Shrewsbury, who had afforded the most material assistance in the publication of many of my theological works, without which, the sale being inconsiderable, I should not have been able to publish them at all.

Before I left Birmingham I preached a funeral sermon for my friend, Dr. Price, and another for Mr. Robinson, of Cambridge, who died with us on a visit to preach our annual charity school sermon. I also preached the last annual sermon to the friends of the college at Hackney. All these three sermons were published.

About two years before I left Birmingham, the question about the 'Test Act', was much agitated both in and out of parliament. This, however, was altogether without any concurrence of mine. I only delivered, and published, a sermon, on the 5th of November, 1789, recommending the most peaceable method of pursuing our object. Mr. Madan, however, the most respectable clergyman in the town, preaching and publishing a very inflammatory sermon on the subject, inveighing in the bitterest manner against the Dissenters in general, and myself in particular, I addressed a number of 'Familiar Letters to the Inhabitants of Birmingham', in our defence. This produced a reply from him, and other letters from me. All mine were written in an ironical and rather a pleasant manner, and in some of the last of them I introduced a farther reply to Mr. Burn, another clergyman in Birmingham, who had addressed to me 'Letters on the Infallibility of the Testimony of the Apostles, concerning the Person of Christ', after replying to his first set of letters in a separate publication.

From these small pieces, I was far from expecting any serious consequences. But the dissenters in general being very obnoxious to the court, and it being imagined, though without any reason, that I had been the chief promoter of the measures which gave them offence, the clergy, not only in Birmingham, but through all England, seemed to make it their business, by writing in the public papers, by preaching and other methods, to inflame the minds of the people against me. And on occasion of the celebration of the anniversary of the French Revolution, on July 14, 1791, by several of my friends, but with which I had little to do, a mob encouraged by some persons in power, first burned the meeting-house in which I preached, then another meeting-house in the town, and then my dwelling-house, demolishing my library, apparatus, and, as far as they could, every thing belonging to me. They also burned, or much damaged, the houses of many dissenters, chiefly my friends; the particulars of which I need not recite, as they will be found in two 'Appeals', which I published on the subject, written presently after the riots.

Being in some personal danger on this occasion, I went to London; and so violent was the spirit of party which then prevailed, that I believe I could hardly have been safe in any other place.

There, however, I was perfectly so, though I continued to be an object of troublesome attention until I left the country altogether. It shewed no small degree of courage and friendship in Mr. William Vaughan, to receive me into his house, and also in Mr. Salte, with whom I spent a month at Tottenham. But it shewed more in Dr. Price's congregation, at Hackney, to invite me to succeed him, which they did, though not unanimously, some time after my arrival in London.

In this situation I found myself as happy as I had been at Birmingham; and contrary to general expectation, I opened my lectures to young persons with great success, being attended by many from London; and though I lost some of the hearers, I left the congregation in a better situation than that in which I found it.

On the whole, I spent my time even more happily at Hackney than ever I had done before; having every advantage for my philosophical and theological studies, in some respect superior to what I had enjoyed at Birmingham, especially from my easy access to Mr. Lindsey, and my frequent intercourse with Mr. Belsham, professor of divinity in the new College, near which I lived. Never, on this side the grave, do I expect to enjoy myself so much as I did by the fire-side of Mr. Lindsey, conversing with him and Mrs. Lindsey on theological and other subjects, or in my frequent walks with Mr. Belsham, whose views of most important subjects were, like Mr. Lindsey's the same as my own.

I found, however, my society much restricted with respect to my philosophical acquaintance; most of the members of the Royal Society shunning me on account of my religious or political opinions, so that I at length withdrew myself from them, and gave my reasons for doing so in the preface to my 'Observations and Experiments on the Generation of Air from Water', which I published in Hackney. For, with the assistance of my friends, I had in a great measure replaced my apparatus, and had resumed my experiments, though after the loss of nearly two years.

Living in the neighbourhood of the New College, I voluntarily undertook to deliver the lectures to the pupils on the subject of 'History and General Policy', which I had composed at Warrington, and also on 'Experimental Philosophy and Chemistry', the 'Heads' of which I drew up for this purpose, and afterwards published. In being useful to this institution, I found a source of considerable satisfaction to myself. Indeed, I have always had a high degree of enjoyment in lecturing to young persons, though more on theological subjects than on any other.

After the riots in Birmingham, I wrote 'An Appeal to the Public' on the subject, and that being replied to by the clergy of the place, I wrote a 'Second Part', to which, though they had pledged themselves to do it, they made no reply; so that, in fact, the criminality of the magistrates, and other principal high-churchmen, at Birmingham, in promoting the riot, remains acknowledged. Indeed, many circumstances which have appeared since that time, shew that the friends of the court, if not the prime ministers themselves, were the favourers of that riot; having, no doubt, thought to intimidate the friends of liberty by the measure.

To my appeal I subjoined various 'Addresses' that were sent to me from several descriptions of persons in England, and abroad; and from them I will not deny that I received much satisfaction, as it appeared that the friends of liberty, civil and religious, were of opinion that I was a sufferer in that cause. From France I received a considerable number of addresses; and when the present *National Convention* was called, I was invited by many of the departments to be a member of it. But I thought myself more usefully employed at home, and that I was but ill qualified for a business which required knowledge which none but a native of the country could possess; and therefore declined the honour that was proposed to me.

But no addresses gave me so much satisfaction as those from my late congregation, and especially of the young persons belonging to it, who had attended my lectures. They are a standing testimony of the zeal and fidelity with which I did my duty with respect to them, and which I value highly.

Besides congratulatory addresses, I received much pecuniary assistance from various persons, and bodies of men, which more than compensated for my pecuniary losses, though what was awarded me at the assizes fell two thousand pounds short of them. But my brother-in-law, Mr. John Wilkinson from whom I had not at that time any expectation, in consequence of my son's leaving his employment, was the most generous on the occasion. Without any solicitation he immediately sent me five hundred pounds, which he had deposited in the French funds, and until that be productive, he allows me two hundred pounds per annum.

After the riots, I published my 'Letters to the Swedenborgian Society', which I had composed and prepared for the press just before.

Mr. Wakefield living in the neighbourhood of the college, and publishing at this time his objections to *public worship*, they made a great impression on many of our young men, and in his preface he reflected much on the character of Dr. Price. On both those accounts I thought myself called upon to reply to him, which I did in a series of 'Letters to a Young Man'. Though he made several angry replies, I never noticed any of them. In this situation I also answered Mr. Evanson's 'Observations on the dissonance of the Evangelists, in a Second Set of Letters to a Young Man'. He also replied to me, but I was satisfied with what I had done, and did not continue the controversy.

Besides the 'Sermon' which I delivered on my acceptance of the invitation to the meeting at Hackney, in the preface to which I gave a detailed account of my *systems of catechizing*, I published two 'Fast Sermons', for the years 1793 and 1794, in the latter of which I gave my ideas of ancient prophecies, compared with the then state of Europe, and in the preface to it

I gave an account of my reasons for leaving the country. I also published a 'Farewell Sermon'.

But the most important of my publications in this situation, were a series of 'Letters to the Philosophers and Politicians of France, on the Subject of Religion'. I thought that the light in which I then stood in that country, gave me some advantage in my attempts to enforce the evidence of natural and revealed religion. I also published a set of 'Sermons on the Evidences of Revelation', which I first delivered by public notice, and the delivery of which was attended by great numbers. They were printed just before I left England.

As the reasons for this step in my conduct are given at large in the preface to my 'Fast Sermon', I shall not dwell upon them here. The bigotry of the country in general made it impossible for me to place my sons in it to any advantage. William had been some time in France, and on the breaking out of the troubles in that country, he had embarked for America, where his two brothers met him. My own situation, if not hazardous, was become unpleasant, so that I thought my removal would be of more service to the cause of truth than my longer stay in England. At length, therefore, with the approbation of all my friends, without exception, but with great reluctance on my own part, I came to that resolution; I being at a time of life in which I could not expect much satisfaction as to friends and society, comparable to that which I left, in which the resumption of my philosophical pursuits must be attended with great disadvantage, and in which success, in my still more favourite pursuit, the propagation of Unitarianism, was still more uncertain. It was also painful to me to leave my daughter, Mr. Finch having the greatest aversion to leave his relations and friends in England.

At the time of my leaving England, my son, in conjunction with Mr. Cooper and other emigrants, had a scheme for a large settlement for the friends of liberty in general, near the head of Susquehanna, in Pennsylvania. And taking it for granted that it would be carried into effect, after landing at New York I went to Philadelphia, and thence came to Northumberland, a town the nearest to the proposed settlement, thinking to reside there until some progress had been made in it. The settlement was given up; but being here, and my wife and myself liking the place, I have determined to take up my residence here, though subject to many disadvantages. Philadelphia was excessively expensive, and this comparatively a cheap place; and my sons settling in the neighbourhood will be less exposed to temptation, and more likely to form habits of sobriety and industry. They will also be settled at much less expence than in or near a large town. We hope, after some time, to be joined by a few of our friends from England, that a readier communication will be opened with Philadelphia, and that the place will improve, and become more eligible in other respects.

When I was at sea, I wrote some 'Observations on the Cause of the present Prevalence of Infidelity', which I published, and prefixed to a new edition of the 'Letters to the Philosophers and Politicians of France'. I have also published my 'Fast and Farewell Sermons', and my 'Small Tracts', in defence of Unitarianism; also a 'Continuation of those Letters', and a 'Third Part of Letters to a Philosophical Unbeliever', in answer to Mr. Paine's 'Age of Reason'.

The observations on the prevalence of infidelity I have much enlarged, and intend soon to print; but I am chiefly employed on the continuation of 'History of the Christian Church'.

Northumberland, March 24, 1795, in which I have completed the sixty-second year of my age.

15.4 Antoine Lavoisier, *Elements of Chemistry*, 1789, trans. R. Kerr, 1790 (London: Dover reprint, 1965), pp. 32–47

CHAP. III.

Analysis of Atmospheric Air, and its Division into two Elastic Fluids; the one fit for Respiration, the other incapable of being respired.

FROM what has been premised, it follows, that our atmosphere is composed of a mixture of every substance capable of retaining the gasseous or aëriform state in the common temperature, and under the usual pressure which it experiences. These fluids constitute a mass, in some measure homogeneous, extending from the surface of the earth to the greatest height hitherto attained, of which the density continually decreases in the inverse ratio of the superincumbent weight. But, as I have before observed, it is possible that this first stratum is surmounted by several others consisting of very different fluids.

Our business, in this place, is to endeavour to determine, by experiments, the nature of the elastic fluids which compose the interior stratum of air which we inhabit. Modern chemistry has made great advances in this research; and it will appear by the following details that the analysis of atmospherical air has been more rigorously determined than that of any other substance of the class. Chemistry affords two general methods of determining the constituent principles of bodies, the method of analysis, and that of synthesis. When, for instance, by combining water with alkohol, we form the species of liquor called, in commercial language, brandy or spirit of wine, we certainly have a right to conclude, that brandy, or spirit of wine, is composed of alkohol combined with water. We can produce the same result by the analytical method; and in general it ought to be considered as a principle in chemical science, never to rest satisfied without both these species of proofs.

We have this advantage in the analysis of atmospherical air, being able both to decompound it, and to form it a new in the most satisfactory manner.

I shall, however, at present confine myself to recount such experiments as are most conclusive upon this head; and I may consider most of these as my own, having either first invented them, or having repeated those of others, with the intention of analysing atmospherical air, in perfectly new points of view.

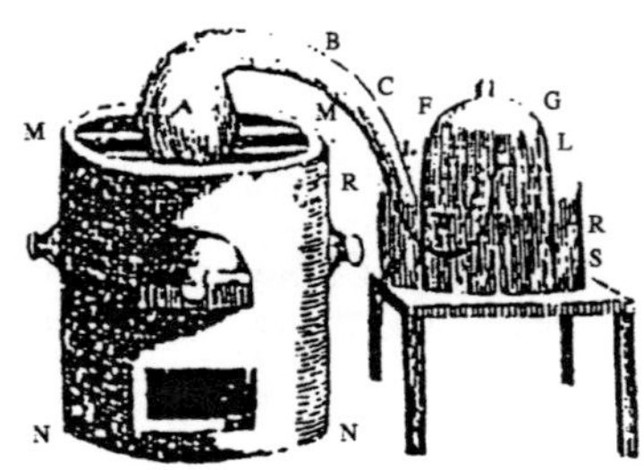

I took a matrass of about 36 cubical inches capacity, having a long neck B C D E, of six or seven lines internal diameter, and having bent the neck as in [the] Plate [above] so as to allow of its being placed in the furnace M M N N, in such a manner that the extremity of its neck E might be inserted under a bell-glass F G, placed in a trough of quicksilver R R S S; I introduced four ounces of pure mercury into the matrass, and, by means of a syphon, exhausted the air in the receiver F G, so as to raise the quicksilver to L L, and I carefully marked the height at which it stood by pasting on a slip of paper. Having accurately noted the height of the thermometer and barometer, I lighted a fire in the furnace M M N N, which I kept up almost continually during twelve days, so as to keep the quicksilver always almost at its boiling point. Nothing remarkable took place during the first day: The Mercury, though not boiling, was continually evaporating, and covered the interior furnace of the vessels with small drops, at first very minute, which gradually augmenting to a sufficient size, fell back into the mass at the bottom of the vessel. On the second day, small red particles began to appear on the surface of the mercury, which, during the four or five following days, gradually increased in size and number; after which they ceased to increase in either respect. At the end of twelve days, seeing that the calcination of the mercury did not at all increase, I extinguished the fire, and allowed the vessels to cool. The bulk of air in the body and neck of the matrass, and in the bell-glass, reduced to a medium of 28 inches of the barometer and 10° (54.5°) of the thermometer, at the commencement of the experiment was about 50 cubical inches. At the end of the experiment the remaining air, reduced to the same medium pressure and temperature, was only between 4[illegible] 43 cubical inches; consequently it had lost about $\frac{1}{6}$ of its bulk. Afterwards, having collected all the red particles, formed during the experiment, from the running mercury in which they floated, I found these to amount to 45 grains.

I was obliged to repeat this experiment several times, as it is difficult in one experiment both to preserve the whole air upon which we operate, and to collect the whole of the red particles, or calx of mercury, which is formed during the calcination. It will often happen in the sequel, that I shall, in this manner, give in one detail the results of two or three experiments of the same nature.

The air which remained after the calcination of the mercury in this experiment, and which was reduced to $\frac{5}{6}$ of its former bulk, was no longer fit either for respiration or for combustion; animals being introduced into it were suffocated in a few seconds, and when a taper was plunged into it, it was extinguished as if it had been immersed into water.

In the next place, I took the 45 grains of red matter formed during this experiment, which I put into a small glass retort, having a proper apparatus for receiving such liquid, or gasseous product, as might be extracted: Having applied a fire to the retort in a furnace, I observed that, in proportion as the red matter became heated, the intensity of its colour augmented. When the retort was almost red hot, the red matter began gradually to decrease in bulk, and in a few minutes after it disappeared altogether; at the same time $41\frac{1}{2}$ grains of running mercury were collected in the recipient, and 7 or 8 cubical inches of elastic fluid, greatly more capable of supporting both respiration and combustion than atmospherical air, were collected in the bell-glass.

A part of this air being put into a glass tube of about an inch diameter, showed the following properties: A taper burned in it with a dazzling splendour, and charcoal, instead of consuming quietly as it does in common air, burnt with a flame, attended with a decrepitating noise, like phosphorus, and threw out such a brilliant light that the eyes could hardly endure it. This species of air was discovered almost at the same time by Mr. Priestley, Mr. Scheele, and myself. Mr. Priestley gave it the name of *dephlogisticated air*, Mr. Scheele called it *empyreal air*. At first I named it *highly respirable air*, to which has since been substituted the term of *vital air*. We shall presently see what we ought to think of these denominations.

In reflecting upon the circumstances of this experiment, we readily perceive, that the mercury, during its calcination, absorbs the salubrious and respirable part of the air, or, to speak more strictly, the base of this respirable part; that the remaining air is a species of mephitis, incapable of supporting combustion or respiration; and consequently that atmospheric air is composed of two elastic fluids of different and opposite qualities. As a proof of this important truth, if we recombine these two elastic fluids, which we have separately obtained in the above experiment, viz. the 42 cubical inches of mephitis, with the 8 cubical inches of respirable air, we reproduce an air precisely similar to that of the atmosphere, and possessing nearly the same power of supporting combustion and respiration, and of contributing to the calcination of metals.

Although this experiment furnishes us with a very simple means of obtaining the two principal elastic fluids which compose our atmosphere, separate from each other, yet it does not give us an exact idea of the proportion in which these two enter into its composition: For the attraction of mercury to the respirable part of the air, or rather to its base, is not sufficiently strong to overcome all the circumstances which oppose this union. These obstacles are the mutual adhesion of the two constituent parts of the atmosphere for each other, and the elective attraction which unites the base of vital air with caloric; in consequence of these, when the calcination ends, or is at least carried as far as is possible, in a determinate quantity of atmospheric air, there still remains a portion of respirable air united to the mephitis, which the mercury cannot separate. I shall afterwards show, that, at least in our climate, the atmospheric air is composed of respirable and mephitic airs, in the proportion of 27 and 73; [. . .]

I mentioned before, that we have two ways of determining the constituent parts of atmospheric air, the method of analysis, and that by synthesis. The calcination of mercury has furnished us with an example of each of these methods, since, after having robbed the respirable part of its base, by means of the mercury, we have restored it, so as to recompose an air precisely similar to that of the atmosphere. But we can equally accomplish this synthetic composition of atmospheric air, by borrowing the materials of which it is composed from different kingdoms of nature. We shall see hereafter that, when animal substances are dissolved in the nitric acid, a great quantity of gas is disengaged, which extinguishes light, and is unfit for animal respiration, being exactly similar to the noxious or mephitic part of atmospheric air. And, if we take 73 parts, by weight, of this elastic fluid, and mix it with 27 parts of highly respirable air, procured from calcined mercury, we will form an elastic fluid precisely similar to atmospheric air in all its properties.

Chapter Sixteen
Conclusions

16.1 F. Voltaire, *Letters Concerning the English Nation*, Letter XIV, 1734, ed. and intro. N. Cronk (Oxford: Oxford University Press, 1994), pp. 61–6

LETTER XIV.

On Des Cartes *and Sir* Isaac Newton.

A FRENCHMAN who arrives in *London*, will find Philosophy, like every Thing else, very much chang'd there. He had left the World a *plenum*, and he now finds it a *vacuum*. At *Paris* the Universe is seen, compos'd of Vortices of subtile Matter; but nothing like it is seen in *London*. In *France*, 'tis the Pressure of the Moon that causes the Tides; but in *England* 'tis the Sea that gravitates towards the Moon; so that when you think that the Moon should make it Flood with us, those Gentlemen fancy it should be Ebb, which, very unluckily, cannot be prov'd. For to be able to do this, 'tis necessary the Moon and the Tides should have been enquir'd into, at the very instant of the Creation.

You'll observe farther, that the Sun, which in *France* is said to have nothing to do in the Affair, comes in here for very near a quarter of its Assistance. According to your *Cartesians*, every Thing is perform'd by an Impulsion, of which we have very little Notion; and according to Sir *Isaac Newton*, 'tis by an Attraction, the Cause of which is as much unknown to us. At *Paris* you imagine that the Earth is shap'd like a Melon, or of an oblique Figure; at *London* it has an oblate one. A *Cartesian* declares that Light exists in the Air; but a *Newtonian* asserts that it comes from the Sun in six Minutes and a half. The several Operations of your Chymistry are perform'd by Acids, Alkalies and subtile Matter; but Attraction prevails even in Chymistry among the *English*.

The very Essence of Things is totally chang'd. You neither are agreed upon the Definition of the Soul, nor on that of Matter. *Descartes*, as I observ'd in my last, maintains that the Soul is the same Thing with Thought, and Mr. *Locke* has given a pretty good Proof of the contrary.

Descartes asserts farther, that Extension alone constitutes Matter, but Sir *Isaac* adds Solidity to it.

> How furiously contradictory are these Opinions!
> *Non nostrum inter vos tantas componere lites.*
>
> VIRGIL, Eclog. III.
>
> *'Tis not for us to end such great Disputes.*

This famous *Newton*, this Destroyer of the *Cartesian* System, died in *March Anno* 1727. His Countrymen honour'd him in his Life-Time, and interr'd him as tho' he had been a King who had made his People happy.

The *English* read with the highest Satisfaction, and translated into their Tongue, the Elogium of Sir *Isaac Newton*, which Mr. *de Fontenelle*, spoke in the Academy of Sciences. Mr. *de Fontenelle* presides as Judge over Philosophers; and the *English* expected his Decision, as a solemn Declaration of the Superiority of the *English* Philosophy over that of the *French*. But when 'twas found that this Gentleman had compar'd *Des Cartes* to Sir *Isaac*, the whole Royal Society in *London* rose up in Arms. So far from acquiescing with Mr. *Fontenelle*'s Judgment, they criticis'd his Discourse. And even several (who however were not the ablest Philosophers in that Body) were offended at the Comparison; and for no other Reason but because *Des Cartes* was a *Frenchman*.

It must be confess'd that these two great Men differ'd very much in Conduct, in Fortune, and in Philosophy.

Nature had indulg'd *Des Cartes* a shining and strong Imagination, whence he became a very singular Person both in private Life, and in his Manner of Reasoning. This Imagination could not conceal it self even in his philosophical Works, which are every where adorn'd with very shining, ingenious Metaphors and Figures. Nature had almost made him a Poet; and indeed he wrote a Piece of Poetry for the Entertainment of *Christina* Queen of *Sweden*, which however was suppress'd in Honour to his Memory.

He embrac'd a Military Life for some Time, and afterwards becoming a complete Philosopher, he did not think the Passion of Love derogatory to his Character. He had by his Mistress a Daughter call'd *Froncine*, who died young, and was very much regretted by him. Thus he experienc'd every Passion incident to Mankind.

He was a long Time of Opinion, that it would be necessary for him to fly from the Society of his Fellow Creatures, and especially from his native Country, in order to enjoy the Happiness of cultivating his philosophical Studies in full Liberty.

Des Cartes was very right, for his Cotemporaries were not knowing enough to improve and enlighten his Understanding, and were capable of little else than of giving him Uneasiness.

He left *France* purely to go in search of Truth, which was then persecuted by the wretched Philosophy of the Schools. However, he found that Reason was as much disguis'd and deprav'd in the Universities of *Holland*, into which he withdrew, as in his own Country. For at the Time that the *French* condemn'd the only Propositions of his Philosophy which were true, he was persecuted by the pretended Philosophers of *Holland*, who understood him no better; and who, having a nearer View of his Glory, hated his Person the more, so that he was oblig'd to leave *Utrecht*. *Des Cartes* was injuriously accus'd of being an Atheist, the last Refuge of religious Scandal: And he who had employ'd all the Sagacity and Penetration of his Genius, in searching for new Proofs of the Existence of a God, was suspected to believe there was no such Being.

Such a Persecution from all Sides, must necessarily suppose a most exalted Merit as well as a very distinguish'd Reputation, and indeed he possess'd both. Reason at that Time darted a Ray upon the World thro' the Gloom of the Schools, and the Prejudices of popular Superstition. At last his Name spread so universally, that the *French* were desirous of bringing him back into his native Country by Rewards, and accordingly offer'd him an annual Pension of a thousand Crowns. Upon these Hopes *Des Cartes* return'd to *France*; paid the Fees of his Patent, which was sold at that Time, but no Pension was settled upon him. Thus disappointed, he return'd to his Solitude in *North-Holland*, where he again pursued the Study of Philosophy, whilst the great *Galileo*, at fourscore Years of Age, was groaning in the Prisons of the Inquisition, only for having demonstrated the Earth's Motion.

At last *Des Cartes* was snatch'd from the World in the Flower of his Age at *Stockholm*. His Death was owing to a bad Regimen, and he expir'd in the Midst of some *Literati* who were his Enemies, and under the Hands of a Physician to whom he was odious.

The Progress of Sir *Isaac Newton*'s Life was quite different. He liv'd happy, and very much honour'd in his native Country, to the Age of fourscore and five Years.

'Twas his peculiar Felicity, not only to be born in a Country of Liberty, but in an Age when all scholastic Impertinencies were banish'd from the World. Reason alone was cultivated, and Mankind cou'd only be his Pupil, not his Enemy.

One very singular Difference in the Lives of these two great Men is, that Sir *Isaac*, during the long Course of Years he enjoy'd was never sensible to any Passion, was not subject to the common Frailties of Mankind, nor ever had any Commerce with Women; a Circumstance which was assur'd me by the Physician and Surgeon who attended him in his last Moments.

We may admire Sir *Isaac Newton* on this Occasion, but then we must not censure *Des Cartes*.

The Opinion that generally prevails in *England* with regard to these two Philosophers is, that the latter was a Dreamer, and the former a Sage.

Very few People in *England* read *Descartes*, whose Works indeed are now useless. On the other Side, but a small Number peruse those of Sir *Isaac*, because to do this the Student must be deeply skill'd in the Mathematicks, otherwise those Works will be unintelligible to him. But notwithstanding this, these great Men are the Subject of every One's Discourse. Sir *Isaac Newton* is allow'd every Advantage, whilst *Des Cartes* is not indulg'd a single one. According to some, 'tis to the former that we owe the Discovery of a *Vacuum*, that the Air is a heavy Body, and the Invention of Telescopes. In a Word, Sir *Isaac Newton* is here as the *Hercules* of fabulous Story, to whom the Ignorant ascrib'd all the Feats of ancient Heroes.

In a Critique that was made in *London* on Mr. *de Fontenelle*'s Discourse, the Writer presum'd to assert that *Des Cartes* was not a great Geometrician. Those who make such a Declaration may justly be reproach'd with flying in their Master's Face. *Des Cartes* extended the Limits of Geometry as far beyond the Place where he found them, as Sir *Isaac* did after him. The former first taught the Method of expressing Curves by Equations. This Geometry which, Thanks to him for it, is now grown common, was so abstruse in his Time, that not so much as one Professor would undertake to explain it; and *Schotten* in *Holland*, and *Format* in *France*, were the only Men who understood it.

He applied this geometrical and inventive Genius to Dioptricks, which, when treated of by him, became a new Art. And if he was mistaken in some Things, the Reason of that is, a Man who discovers a new Tract of Land cannot at once know all the Properties of the Soil. Those who come after him, and make these Lands fruitful, are at least oblig'd to him for the Discovery. I will not deny but that there are innumerable Errors in the rest of *Des Cartes*'s Works.

Geometry was a Guide he himself had in some Measure fashion'd, which would have conducted him safely thro' the several Paths of natural Philosophy. Nevertheless he at last abandon'd this Guide, and gave entirely into the Humour of forming Hypotheses; and then Philosophy was no more than an ingenious Romance, fit only to amuse the Ignorant. He was mistaken in the Nature of the Soul, in the Proofs of the Existence of a God, in Matter, in the Laws of Motion, and in the Nature of Light. He admitted innate Ideas, he invented new Elements, he created a World; he made Man according to his own Fancy; and 'tis justly said, that the Man of *Des Cartes* is in Fact that of *Des Cartes* only, very different from the real one.

He push'd his metaphysical Errors so far, as to declare that two and two make four, for no other Reason but because God would have it so. However, 'twill not be making him too great a Compliment if we affirm that he was valuable even in his Mistakes. He deceiv'd himself, but then it was at least in a methodical Way. He destroy'd all the absurd Chimæra's with which Youth had been infatuated for two thousand Years. He taught his Cotemporaries how to reason, and enabled them to employ his own Weapons against himself. If *Des Cartes* did not pay in good Money, he however did great Service in crying down that of a base Alloy.

I indeed believe, that very few will presume to compare his Philosophy in any respect with that of Sir *Isaac Newton*. The former is an Essay, the latter a Master-Piece: But then the Man who first brought us to the Path of Truth, was perhaps as great a Genius as he who afterwards conducted us through it.

Des Cartes gave Sight to the Blind. These saw the Errors of Antiquity and of the Sciences. The Path he struck out is since become boundless. *Rohault*'s little Work was during some Years a complete System of Physicks; but now all the Transactions of the several Academies in *Europe* put together do not form so much as the Beginning of a System. In fathoming this Abyss no Bottom has been found. We are now to examine what Discoveries Sir *Isaac Newton* has made in it.

16.2 Bernard le Bovier de Fontenelle, *Conversations on the Plurality of Worlds*, 1686, 1687 ('The Sixth Evening')

The Sixth Evening.

[...] I see what is, and always will be, the reason, why the opinion of the planets being inhabited, is not thought so probable as it really is: the planets always present themselves to our view as bodies which emit light; and not at all like great plains and meadows. We should readily agree that plains and meadows were inhabited; but for luminous bodies to be so too, there is no ground to believe it. Reason may come and tell us over and over, that there are plains and meadows in these planets, but reason comes a day too late; one glance of our eyes has had its effect before her; we will not hear a word she says, the planets must be luminous bodies, and what sort of inhabitants should they have, our imagination of course would presently represent their figures to us. It is what she cannot do, and the shortest way is to believe there are no such beings. Would you have me, for the establishment of these planetary people, whose interests are far from touching me, go to attack those formidable powers, called sense and imagination? It is an enterprize [which] would require a good stock of courage, and we cannot easily prevail on men to substitute their reason in the place of their eyes. I sometimes meet with reasonable people enow, who are willing, after a thousand demonstrations, to believe that the planets are so many earths: but their belief is not such as it would be, if they had not seen them under a different appearance; they still remember the first idea they entertained, and they cannot well recover themselves from it. It is these kind of people, who, in believing our opinion, seem to do it a courtesy, and only favour it for the sake of a certain pleasure which its singularity gives them.

WELL, says the marchioness, interrupting me, and is not this sufficient for an opinion, which is but barely probable? You would be very much surprized, says I, if I should tell you, probable is a very modest term. Is it simply probable that there ever was such a man as Alexander the great? You hold it very certain that there was, and upon what is this certainty founded? Because you have all the proofs which you could desire in a like matter; and there does not the least subject for doubt present itself, to suspend or arrest your determination: for you never could see this Alexander, and you have not one mathematical demonstration that there ever was such a man. Now what would you say if the inhabitants of the planets were almost in the very same case? We cannot pretend to make you see them, and you cannot insist upon the demonstration here, as you would in a mathematical question; but you have all the proofs you could desire in our world. The entire resemblance of the planets with the earth which is inhabited, the impossibility of conceiving any other use for which they were created, the fecundity and magnificence of nature, the certain regards she seems to have had to the necessities of their inhabitants, as in giving moons to those planets remote from the sun, and more moons still to those yet more remote; and what is still very material, there are all things to be said on one side, and nothing on the other; and you cannot comprehend the least subject for a doubt, unless you will take the eyes and understanding of the vulgar. In short, supposing that these inhabitants of the planets really exist, they could not declare themselves by more marks, or marks more sensible; and after this you are to consider whether you are willing not to take their case to be more than purely probable. But you would not have me, says she, look upon this to be as certain as that there was such a man as Alexander? Not altogether, madam, says I, for though we have as many proofs touching the inhabitants of the planets, as we can have, in the situation we are, yet the number of these proofs is not great. I must renounce these planetary inhabitants, said her ladyship, interrupting me, for I cannot conceive how to rank them in my imagination; there is no absolute certainty of them, and yet there is more than a probability; so that I am confounded in my notions. Ah, madam, says I, never put yourself out of conceit with them for that; the most common and ordinary clocks shew the hours, but those are wrought with more art and nicety which shew the minutes. Just so your ordinary capacities are sensible of the difference betwixt a simple probability, and an evident certainty; but it is only your fine spirits that discern the exact proportions of certainty or probability, and can mark, if I may use the phrase, the minutes in their sentiments. [...]

...What a notable spot might the lunar-inhabitants all of a sudden discover on our earth; for you know, madam, that seas are spots. It is no less than the common opinion, that Sicily was separated from Italy, and Cyprus from Syria: There are sometimes new islands formed in the seas: earthquakes have swallowed up mountains, others have rose and altered the course of the planets. The philosophers give us apprehensions, that the kingdoms of Naples and Sicily, which are countries founded upon great subterranean vaults full of sulphur, will one day sink in, when those vaults shall no longer be able to resist the flames which they contain, and at this time exhale at those vent-holes, the mouths of Vesuvius and Etna. Is not here enough to diversify the sight which we give to the people in the moon?

[...] I conceive that the sun may be veiled by nature, to be more proportioned to our use. Well, madam, replied I, this is some small introduction to a system which you have very happily started. We may add, that these vapours produce a kind of rain, which falling back upon the sun, may cool and refresh it, as we sometimes throw water into a forge, when the fire is too fierce. There is not any thing but what we may imagine, to assist nature's address, but she has another kind of address very particular, which is to conceal herself from us, and we should not willingly be confident that we have found out her method of acting on her designs in it: in case of new discoveries, we should not be too importunate in our reasonings, though we are always fond enough to do it; and your true philosophers are like elephants, who as they go, never put their second foot to the ground, till their first be well fixed. The comparison seems the more rational to me, says she, as the merit of those two species of animals, elephants and philosophers, does not at all consist in exterior agreements. I am willing to mistake the jugement of both; now teach me some of the latter discoveries, and I promise you not to make any rash systems.

I will tell you, madam, replied I, all the news I know from the firmament, and I believe the freshest advices you can have. I am sorry they are not as surprizing and wonderful, as some observations which I read the other day in an abrigement of the Chinese Annals, written in Latin. Those people see thousands of stars, at a time, fall from the sky into the sea, with a prodigious noise, or are dissolved, and melt into rains; and these are things which have been seen more than once in China. I met with this observation at two several times, pretty distant from each other, without reckoning a certain star which goes eastward, and bursts like a squib, always with a great noise. It is great pity that these kinds of Phenomena should be reserved for China only, and that our part of the globe should never have their share of these sights. It is not long, since all our philosophers were of opinion, that they might affirm on good grounds, that the heavens and all the celestial bodies were incorruptible, and therefore incapable of change; and yet at the same time, there were some men in the other part of the earth who saw stars dissolve by thousands, which must produce a very different opinion. But, says the marchioness, did we ever hear it allowed that the Chinese were such great astronomers? It is true, we did not, says I, but the Chinese have an advantage from being divided from us by such a prodigious tract of earth, as the Greeks had over the Romans, by being so much prior in time: distances of every sort pretend a right of imposing on us. In reality, I think still more and more, that there is a certain genius which has never yet been out of the limits of Europe, or at least not much beyond them; perhaps he may not be permitted to spread over any great extent of the earth at once, and that some fatality prescribes him very narrow bounds. Let us indulge him whilst we have him; the best of it is, he is not linked to the sciences and dry speculations, but launches out with as much success into subjects of pleasure, in which point I question whether any people equal us. These are such topics, madam, as ought to give you entertainment, and compleat your whole system of philosophy.

16.3 Frederick II, *Discourse on the Usefulness of the Arts and Sciences in a State*, 27 January 1772, in *Oeuvres de Frederic le Grand*, vol. IX, repr. and trans. in *The Enlightenment*, Texts I, ed. S. Eliot and K. Whitlock (Milton Keynes: Open University Press, 1992), pp. 66–7

Some unenlightened or hypocritical persons have ventured to profess their hostility to the arts and sciences.[1] If they have been allowed to slander that which does most honour to humanity, we must be all the more entitled to defend it, for that is the duty of all who love society and who are grateful for what they owe to literature. Unfortunately, paradox often makes a greater impression on the public than truth; it is then that we must disabuse the public and refute the authors of such nonsense, not with insults but with sound reason. I am ashamed to state in this Academy that people have had the effrontery to ask whether the sciences are useful or harmful to society, a subject on which no one should entertain the slightest doubt. If we have any superiority over animals, it is certainly not in our bodily faculties, but in the greater understanding which nature has given us; and what distinguishes one man from another is genius and learning. Where lies the infinite difference between a civilized people and barbarians if not in the fact that the former are enlightened, while the others vegetate in brutish ignorance?

The nations which have enjoyed this superiority have been grateful to those who brought them this advantage. Hence the reputation justly enjoyed by those great thinkers of the world, those sages, who through their learned works, have enlightened their compatriots and their age.

Man in himself is little enough; he is born with faculties more or less ripe for development. But they require cultivation; his knowledge must increase if his ideas are to broaden; his memory must be filled if it is to supply the imagination with material on which to work, and his judgement must be refined if it is to discriminate between its own products. The greatest mind, without knowledge, is only a rough diamond that will acquire value only after it has been cut by the hands of a skilled jeweller. What minds have been thus lost to society, what great men of every kind stifled in the bud, whether through ignorance, or through the abject state in which they found themselves placed!

The true benefit of the State, its advantage and glory demand therefore that the people in it should be as well educated and enlightened as possible,

---

1 Rousseau in his *Discourse on the Arts and Sciences* (1750).

in order to furnish it, in every field, with a number of trained subjects capable of acquitting themselves expertly in the different tasks entrusted to them. There are some false political thinkers, limited by their narrow ideas, who, without going deeply into the matter, have supposed that it is easier to rule an ignorant and stupid people than an enlightened nation. That is a really powerful argument, when experience proves that the more brutish the people are, the more they are capricious and obstinate, and that it is far more difficult to overcome its obstinacy than to explain realities to a people civilized enough to listen to reason! It would be a fine country where talents remained forever stifled and where there was only one man less narrow-minded than the rest! Such a State, inhabited by ignoramuses, would be like the lost paradise in Genesis, inhabited only by animals.

Although it is unnecessary to demonstrate to this illustrious audience and in this Academy that the arts and sciences bring both utility and fame to the peoples who possess them, it will perhaps not be without use to convince some less enlightened persons of the same thing, to arm them against the effects which some vile sophists might have on their minds. Let them compare a Canadian savage with any citizen of a civilized country of Europe, and all the advantage will be with the latter. How can one prefer crude nature to nature perfected, lack of means of subsistence to a life of ease, rudeness to politeness, the security of possessions enjoyed under the protection of the laws to the law of the jungle and to anarchy, which destroys the fortunes and conditions of families?

Society, a community of men, could not do without either the arts or the sciences. Thanks to surveying and hydraulics, riparian regions are protected from flooding; without these arts, fertile lands would become unhealthy marshes, and would deprive numerous families of their livelihood. The higher lands could not do without surveyors to measure out and divide the fields. The physical sciences, firmly established by experiment, help to perfect agriculture and, in particular, horticulture. Botany, applied to the study of medicinal herbs, and chemistry, which can extract their essences, serve at least to fortify our hope during our illnesses, even if their property cannot cure us. Anatomy guides and directs the surgeon's hand in those painful but necessary operations that save our life at the expense of an amputated limb.

The mechanical sciences are useful in every field: if a load is to be raised or transported, they will move it. If we are to dig into the bowels of the earth to extract metals, the science of mechanics, with ingenious machines, pumps out the quarries and frees the miner from the super-abundance of water which would cost him his life or his work. If we need mills to grind the most familiar and most basic form of food, the science of mechanics perfects them. It is the science of mechanics that helps craftsmen by improving the various kinds of craft at which they work. Every kind of machine lies within its province. And how many machines are needed in all the various fields! The craft of shipbuilding constitutes perhaps one of the greatest efforts of imagination; but how much knowledge the pilot must possess to steer his ship and brave wind and wave! He needs to have studied astronomy, to have good charts, an exact knowledge of geography and arithmetical skill, in order to ascertain the distance he has travelled and the point at which he is, and in this respect he will be helped in future by the chronometers which have just been perfected in England.[2] The arts and sciences go hand in hand: we owe them everything, they are the benefactors of mankind.

We agree that logic is beyond the riff-raff; this large section of the human race will always be the last to open its eyes to facts; but although in every country, it maintains a store of superstitions, it is also true to say that we have succeeded in disabusing it of its belief in witches, possession by devils, the philosopher's stone and other equally childish nonsense. We owe these advantages to the more meticulous study of nature which has been carried out. Physics has been combined with analysis and experiment; the brightest light has been brought to bear among these dark places which concealed so many truths from scholars of the past; and although we cannot attain to the knowledge of the first secret principles, which the Grand Geometrician[3] has reserved for himself alone, powerful geniuses have arisen who discovered the laws of gravity and motion. Chancellor Bacon, the precursor of the new philosophy, or rather, the man who guessed and predicted its progress, put Sir Isaac Newton on the track of his marvellous discoveries; Newton appeared after Descartes, who having discredited the errors of the past, replaced them with errors of his own. Since then, men have weighed the air, measured the skies, calculated the movement of the heavenly bodies with infinite accuracy, predicted eclipses, and discovered an unknown property of matter, the force of electricity, whose effects astound the imagination; and doubtless men will soon be able to predict the appearance of comets, as they do that of eclipses, (though we already owe to the learned Bayle[4] the dissipation of the fear which this phenomenon caused among the ignorant). Let us admit: while the weakness of our condition makes us humble, the works of these great men restore our courage and make us feel the dignity of our being.

---

2 John Harrison's chronometers, tested by Captain Cook during his second voyage.

3 i.e. God. The expression 'Grand Geometrician' was used by Freemasons.

4 In his *Miscellaneous Thoughts on the comet of 1680* (1682), 'wherein it is proved by several reasons drawn from philosophy and theology that comets are not presages of misfortune'.

Index

Note: Extracts from works are indicated by the use of **bold** type. n = footnote.

CPSIA information can be obtained at www.ICGtesting.com
Printed in the USA
BVOW041510231111

276705BV00005B/2/P